Unlocking Salesforce Data

A Complete Guide to Designing Secure and Scalable Sharing & Visibility Architecture

Sandhya Sharma

Apress®

Unlocking Salesforce Data: A Complete Guide to Designing Secure and Scalable Sharing & Visibility Architecture

Sandhya Sharma
Hyderabad, Telangana, India

ISBN-13 (pbk): 979-8-8688-2304-6
https://doi.org/10.1007/979-8-8688-2305-3

ISBN-13 (electronic): 979-8-8688-2305-3

Copyright © 2026 by Sandhya Sharma

This work is subject to copyright. All rights are reserved by the Publisher, whether the whole or part of the material is concerned, specifically the rights of translation, reprinting, reuse of illustrations, recitation, broadcasting, reproduction on microfilms or in any other physical way, and transmission or information storage and retrieval, electronic adaptation, computer software, or by similar or dissimilar methodology now known or hereafter developed.

Trademarked names, logos, and images may appear in this book. Rather than use a trademark symbol with every occurrence of a trademarked name, logo, or image we use the names, logos, and images only in an editorial fashion and to the benefit of the trademark owner, with no intention of infringement of the trademark.

The use in this publication of trade names, trademarks, service marks, and similar terms, even if they are not identified as such, is not to be taken as an expression of opinion as to whether or not they are subject to proprietary rights.

While the advice and information in this book are believed to be true and accurate at the date of publication, neither the authors nor the editors nor the publisher can accept any legal responsibility for any errors or omissions that may be made. The publisher makes no warranty, express or implied, with respect to the material contained herein.

Managing Director, Apress Media LLC: Welmoed Spahr
Acquisitions Editor: Aditee Mirashi
Desk Editor: James Markham
Editorial Project Manager: Rachel Zhang

Cover designed by eStudioCalamar

Distributed to the book trade worldwide by Springer Science+Business Media New York, 1 New York Plaza, New York, NY 10004. Phone 1-800-SPRINGER, fax (201) 348-4505, e-mail orders-ny@springer-sbm.com, or visit www.springeronline.com. Apress Media, LLC is a Delaware LLC and the sole member (owner) is Springer Science + Business Media Finance Inc (SSBM Finance Inc). SSBM Finance Inc is a **Delaware** corporation.

For information on translations, please e-mail booktranslations@springernature.com; for reprint, paperback, or audio rights, please e-mail bookpermissions@springernature.com.

Apress titles may be purchased in bulk for academic, corporate, or promotional use. eBook versions and licenses are also available for most titles. For more information, reference our Print and eBook Bulk Sales web page at http://www.apress.com/bulk-sales.

Any source code or other supplementary material referenced by the author in this book is available to readers on GitHub. For more detailed information, please visit https://www.apress.com/gp/services/source-code.

If disposing of this product, please recycle the paper.

Table of Contents

About the Author

 **Sandhya Sharma** is a Solutions Architect, AI Strategy Leader, and Co-Leader of the Salesforce Women in Tech Group in Hyderabad, India. She is a dynamic Salesforce professional with a strong focus on designing scalable, secure, and purpose-driven solutions, currently working at Salesforce.

Though her journey into Salesforce architecture began with a non-technical background, her ability to bridge business context with technical design has influenced her approach to solution architecture—prioritizing clarity, governance, and long-term sustainability over short-term fixes.

Sandhya brings hands-on experience across multiple Salesforce clouds, including Sales, Service, Finance, Field Service, Health Cloud, Agentforce, and Data Cloud. She holds 19 Salesforce certifications, along with MuleSoft certification, two AMP Impact certifications, and a Scale Up Architect certification. Driven by her long-term goal of achieving the Salesforce Certified Technical Architect (CTA) credential, she remains committed to continuous learning and architectural excellence.

A frequent speaker at community events such as MuleDreamin'24 and various Salesforce conferences, Sandhya actively contributes to Trailblazer groups across Hyderabad and India and volunteers with Hyderabad Trailblazin'. She is a strong advocate for continuous upskilling and is particularly interested in the responsible application of AI within enterprise platforms—exploring how intelligence and automation can enhance decision-making without compromising trust, security, or governance.

Outside her professional work, she is a proud mother of two and a passionate supporter of inclusive growth within the Salesforce ecosystem.

About the Technical Reviewer

Sanyam Jain is a globally recognized Security Engineering Architect and cybersecurity thought leader, known for securing complex digital ecosystems and strengthening resilience across cloud-native environments. With expertise in Cloud Security, Security Operations, Application Security, Compliance, and Security Automation, he helps enterprises and startups exceed their security goals through strategic foresight and technical excellence.

Embracing a DevSecOps-first approach, Sanyam designs secure architectures, embeds security in CI/CD pipelines, and automates compliance. His skills span threat detection, Kubernetes security, IAM, encryption, and network security across AWS, Azure, and Google Cloud. He is well-versed in frameworks such as ISO 27001, SOC 2, SOX, HITRUST, GDPR, HIPAA, PCI DSS, and NIST CSF, enabling secure-by-design implementations that meet audit requirements.

Sanyam's research has been featured in *Forbes*, *TechCrunch*, *ZDNet*, and other leading publications. He also serves as a Judge for the Globee Awards and the Business Intelligence Group, an Evaluator in the Smart India Hackathon, a mentor with NITI Aayog, and an advisor to startups and NGOs like the GDI Foundation.

As a reviewer for *O'Reilly*, *Apress*, and *BPB Publications*, he has mentored thousands through Udacity and other learning platforms. Sanyam holds a master's degree in Technology from BITS Pilani and is a Certified Kubernetes Administrator.

Acknowledgments

This book has been shaped not only by my own experiences, but also by the generosity and insight of members of the Salesforce community who took the time to share their perspectives for the section "Tips from Industry Experts on Sharing" in Chapter 6. I would like to extend my sincere thanks to Buyan Thyagarajan, Dave Norris, Dinesh Yadav, Jitendra Zaa, Lilith Van Beisen, Matt Francis, and Walter Spinard, along with one contributor who chose to remain anonymous, for their thoughtful and candid contributions.

Each perspective reflects years of hands-on experience across diverse industries and organizational contexts, bringing practical depth to the discussions in this book. Their willingness to share lessons learned, challenges encountered, and principles they trust in real-world implementations has helped ground this work beyond theory. I am deeply grateful for their support and for the role they played in strengthening both the clarity and relevance of the ideas presented here.

Introduction

Data access is one of the most deceptively complex aspects of any Salesforce implementation. At first glance, sharing and visibility appear straightforward—roles, profiles, permission sets, and a few rules seem sufficient to get users productive. Yet as organizations grow, regulations tighten, teams restructure, and data volumes increase, access decisions made early in an implementation often surface as the hardest problems to untangle later.

This book is about that reality.

Salesforce's security and sharing model is powerful, but it is also deeply interconnected. Small design choices can have long-lasting consequences, affecting performance, compliance, user trust, and operational scalability. What works for a small team rarely survives unchanged in a large enterprise, nonprofit, or regulated environment. Sharing models do not fail loudly; they drift, accumulate assumptions, and eventually become fragile.

The goal of this book is to help you design sharing and visibility models that endure.

Who This Book Is For

This book is written for Salesforce professionals who already understand the basics and want to move beyond configuration into architectural thinking. It is especially relevant for

- Salesforce Administrators looking to deepen their understanding of visibility design

- Developers working with complex access requirements

- Solution Architects responsible for long-term scalability and governance

- Aspiring Salesforce Architects and CTA candidates preparing for real-world scenarios and exams

If you have ever struggled to explain *why* a user can see a record, worried about access during audits, or hesitated to change sharing logic because "it might break something," this book is meant for you.

What This Book Covers

Rather than treating sharing as a checklist of features, this book approaches visibility as an architectural discipline—one that sits at the intersection of people, data, and process.

The early chapters establish a strong foundation, revisiting Salesforce's security layers, profiles, permission sets, role hierarchies, and organization-wide defaults—not to repeat documentation, but to clarify how these pieces interact in practice. From there, the book moves into deeper territory: criteria-based sharing, manual sharing, and Apex-managed sharing, with a strong emphasis on *when* each approach is appropriate and *when it is not*.

Later chapters focus on what happens after go-live: performance considerations, troubleshooting visibility issues, scaling sharing models as organizations grow, and governing access over time. These sections reflect lessons learned from real implementations, including nonprofit, enterprise, regulated, and multi-org environments.

The final chapters shift toward application and assessment. You will find architect-level case studies, scenario-based questions, and exam-style analyses designed to test not just your knowledge of Salesforce features, but your ability to reason through trade-offs, constraints, and long-term impact.

How the Book Is Structured

Each chapter is designed to build on the previous one, moving from foundational concepts to advanced decision-making:

- Chapters 1–3 establish the core principles of Salesforce data security, access control, role hierarchies, and organization-wide defaults.

- Chapters 4 and 5 explore sharing rules, manual sharing, and Apex-managed sharing, with a focus on architectural intent rather than technical novelty.

- Chapter 6 addresses optimization, troubleshooting, and scaling—how sharing models behave under real organizational pressure.
- Chapter 7 brings everything together through case studies, exam-style questions, and practical guidance for certification preparation.

Throughout the book, you will also find brainstorming sections designed to encourage reflection. These are not exercises with right or wrong answers, but prompts meant to help you examine your own implementations with a more critical, architectural lens.

How to Use This Book

This is not a book you need to read cover to cover in one sitting. You may choose to read chapters sequentially, or you may return to specific sections as reference points when designing or reviewing a sharing model.

If you are preparing for an architect-level role or certification, pay close attention to the case studies and scenario-based questions. They are intentionally written to reflect ambiguity, constraints, and imperfect information—because that is how real projects behave.

Most importantly, this book encourages restraint. The most successful sharing models are rarely the most complex. They are the ones that remain understandable, defensible, and adaptable long after the original implementation team has moved on.

A Final Note Before You Begin

Salesforce gives us many tools to control access. The challenge is not knowing *how* to use them, but knowing *why, when,* and *for how long.*

As you read, resist the urge to look for a single "best practice." Instead, focus on patterns, trade-offs, and questions that help you design with intention. Visibility decisions shape trust, compliance, and collaboration—and they deserve the same architectural care as any other part of the platform.

With that, let's begin by grounding ourselves in the fundamentals of Salesforce data security and sharing.

CHAPTER 1

Introduction to Salesforce Data Security and Sharing

Salesforce's Data Security Framework is the necessary framework that organizations seek to maintain both their internal data privacy and their compliance with various regulatory standards such as GDPR (General Data Protection Regulation), HIPAA (Health Insurance Portability and Accountability Act), and CCPA (California Consumer Privacy Act). Since a data breach can seriously affect the financial and reputational status of a business, the industry also requires the use of a sophisticated and adaptive model of data security. In this respect, Salesforce's security architecture has been designed by taking into consideration the assignment of each level of access, defined and enforced based on a business and legal need.

Overview of Salesforce's Data Security Layers

At its core, this framework is designed around a "multi-layered model" that provides businesses with controls ranging from the broadest to the most granular level—Organization-Level Security, Object-Level Security, Field-Level Security, and Record-Level Security—and each of them plays a crucial role in how data access gets managed in the system (Figure 1-1).

© Sandhya Sharma 2026

S. Sharma, *Unlocking Salesforce Data*, https://doi.org/10.1007/979-8-8688-2305-3_1

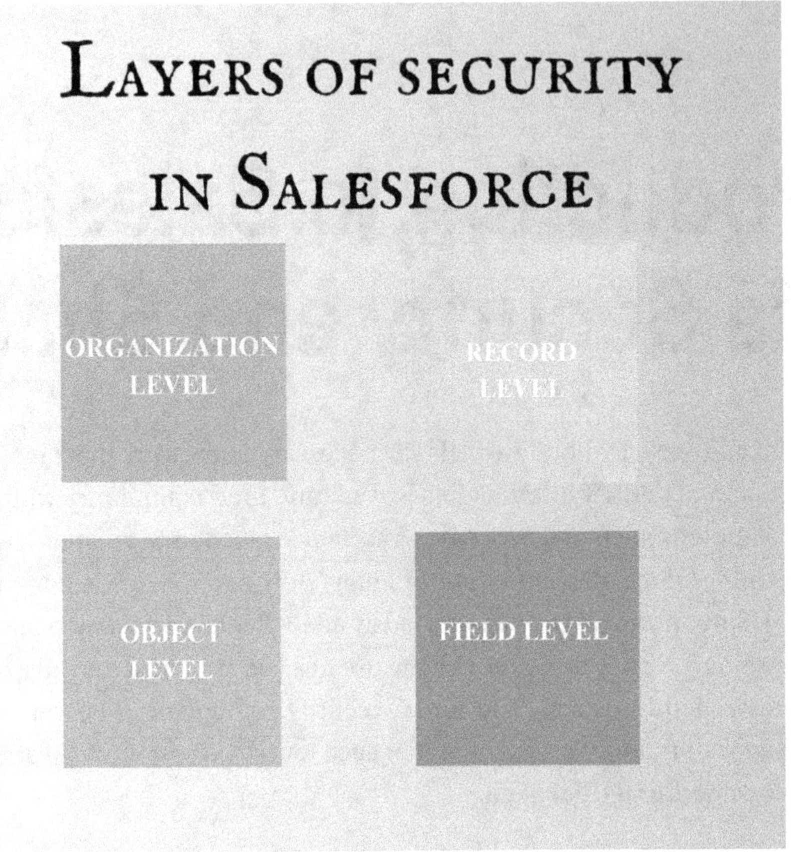

Figure 1-1. *Salesforce security layers*

Breaking down data access into these layers empowers Salesforce to allow the administrator to build a customized setup for data visibility and security configurations that match user responsibilities and needs.

- **Organization-Level Security**: It is the outermost security layer, governing *when, where, and under what conditions* users can access the Salesforce org itself; it includes controls such as login hours, IP range restrictions, network access, and session settings, ensuring that only trusted users in approved contexts can enter the environment— before any object, field, or record access is even evaluated.

- **Object-Level Security**: It is a foundational layer, giving control over broad access to entire data objects, such as Accounts, Contacts, or Cases; it determines whether the user can see, create, edit, or delete records of any given type.

- **Field-Level Security**: A higher layer of security is field level. Field-level security will allow even more granularity, by letting administrators control the level of access at the field level within each object. It is useful in handling very sensitive information that doesn't need to be visible to all users who have access to the object itself.

- **Record-Level Security**: The last layer is Record-Level Security, offering complete control over specific records within an object. Again, this layer is of great importance to share or restrict certain records to specific roles of the organization or even team structures so that businesses can fine-tune access for smooth day-to-day operations.

All these layers get further fortified by the tools provided by Salesforce, which are Role Hierarchies, Sharing Rules, and Organization-Wide Defaults (OWDs) that strengthen the framework by providing exceptions along with customized settings. For instance, role hierarchies automatically provide access to users who are higher up in the organizational structure, and sharing rules allow exceptions in the default sharing model on peculiar criteria.

All these layers feature their tools and options for configuring data visibility and managing how access can be managed in the best manner fitting the needs of any organization. For example, one company might wish to limit access to highly sensitive information about customers only to a small number of team members, while being more open on the general information of accounts across departments. Salesforce's multiple-layered security allows you to achieve this kind of granularity with ease.

Layered Security Is Critical

One security setting rarely suffices when dealing with sensitive data within highly trafficked organizations; the Salesforce model addresses this with the segmentation of data access across distinct layers. Each layer serves a specific purpose within the overall framework, combining to provide a level of data protection while providing flexibility to support business processes and collaboration. This layered approach is meant to be neither too limited in access to data, thereby cutting out much productivity, nor too open, which might eventually introduce possible risks. This flexible model has a way of enabling businesses to adjust their security settings as their operations change, and they can grow and innovate without compromising safety over data.

Flexibility for Complex Business Needs

Salesforce.com offers a layered framework of data security that can be tailored to be very flexible and responsive to the needs of businesses of all sizes and for businesses of all types—from small startups with modest requirements to large corporations subject to stringent regulatory standards. Salesforce strikes this balance between competing priorities such as access vs. protection by allowing administrators to segment visibility at several layers. This is particularly important in complex, matrixed organizations where there may be different users who require varying levels of access based on the roles that they perform, the responsibilities they are charged with, and the nature of work they do.

This multi-layered approach helps in tailoring an organization's data security settings so that it can stop unauthorized access, still allowing collaboration and information sharing that is so important for business success. Each level of Salesforce data security has been designed to be scalable and flexible, so organizations can adapt their settings as their needs for data protection evolve over time.

The rest of the chapters focus on each of these layers in depth, providing more detail on how they work in isolation as well as how they interact and blend together to form a solid, flexible security model. We're going to look at how these layers work individually and together to create a security architecture supporting diverse business needs while keeping the data safe and secure. Whether implementing data access for a small organization or for global enterprises, it means you will understand and implement Salesforce's multi-layered approach to security, and you will truly build a trusted and compliant data environment. This is a high-level overview of the layers of data security; we will take the dive on each layer and how to configure them effectively.

Importance of Share and Visibility for Compliance

The sharing and visibility settings in Salesforce provide the foundation of a safe, compliant environment for the data. Not just an access management tool, they form key parts of an organization's integrity, supporting compliance and the trust it builds among its customers. With thoughtful application of these settings, data can be ensured to reach who needs it to see, and nobody else, all while complying both with internal policies and more general expectations in the wider industry.

Using Controlled Access to Ensure Privacy

For the most part, Salesforce administrators have access to all types of tools that can be implemented in fine detail to customize access levels. Organization-Wide Defaults provide a basic access control layer by setting a minimum level of data visibility across all organization-wide users. While these defaults will establish an equal baseline for all the users, Role Hierarchies and Sharing Rules will typically involve more granular permissions. Customization of these settings allows an organization to balance the measure of accessibility against the measure of security, in which providing necessary access without breaching privacy is feasible.

For example, the sales team of an organization may require read and write access rights to the client accounts but should not be exposed to the financial forecasts or sensitive customer data maintained by another team. OWDs can then be secured at a high level and exceptions are allowed through Role Hierarchies so each team has what is necessary while still safeguarding the data most critical to the organization.

Compliance As an Enabler of Business Agility

There is absolutely no trade-off between the compliance and efficiency requirements because Salesforce's sharing and visibility features are designed to meet both requirements. Administrators can allow efficient access to the right information to the right user with unique roles and responsibilities in a business to enhance productivity.

This makes it possible for Salesforce to automate workflows, eliminate distractions, and help users be more productive. For example, sales reps can close deals, customer service agents can solve cases, and finance teams can work with the records without any extraneous data interfering with them. This also helps avert "data fatigue," by providing users with what they need to perform their roles.

Building Customer Trust Through Transparency and Security

Data security concerns are at its peak these days, and hence any business needs to have confidence-building measures to let the customer feel safe with regard to their information. Companies that take security as a priority will gain more trust from customers; the outcome will have strengthened relationships and long-term loyalty.

Sharing and visibility settings of Salesforce can empower such a commitment, allowing companies to prove a strong privacy-first approach. These access settings can be reviewed and refined periodically, thereby ensuring that each layer of access is updated with the latest privacy commitments, which is very essential in building long-term trust with customers.

Responsive Controls on Access As Compliance Standards Change

As data security standards evolve, Salesforce's sharing model makes it easy to adapt access controls without compromising efficiency. The flexibility of these settings enables organizations to adjust sharing rules, hierarchies, and field-level controls to meet new privacy expectations and keep workflows seamless. Administrators can proactively adjust visibility to maintain compliance and even anticipate emerging data security requirements, all while ensuring the organization's data-sharing model evolves with its growth.

In the following chapters, we'll explore each Salesforce setting and tool that makes this level of data security possible. You'll gain practical insights into configuring a secure, adaptable sharing model that's both compliance-ready and operationally efficient.

Key Components of the Sharing Architecture

The data access of an organization is managed by Salesforce's sharing architecture. It enables administrators to apply significant control over the data visibility. In doing so, Salesforce ensures that the right people have access to the right data without compromising security. At its core, there are a few core elements in the sharing model: Organization-Wide Defaults (OWDs), Role Hierarchies, Sharing Rules, and Manual Sharing. These elements are important for defining a starting point for access and enabling detailed adjustments of record and data sharing in an organization.

These are the different parts that make up the structures of Salesforce's sharing model. But how these are configured and applied in particular use cases for each are discussed in chapters later in the book. The following sections provide an overview of each of these components to set the foundation for a further investigation into them later in the book.

Organization-Wide Defaults

One of the most significant and foundational concepts of the security model in Salesforce is that of Organization-Wide Defaults, or OWDs. OWDs define the base level of access to records for all users in the organization. These settings ensure that, by default, records are either private, open for read-only access, or open for full read/write access. OWDs are important because they enable administrators to lock down access at a record level.

In Salesforce, OWDs can be set for each object, and there are three main options:

- **Private**: Only the record owner and users with special permissions can view or modify the record. It is usually used to maintain privacy of the sensitive information so that it will be accessible only to those people who require access.

- **Public Read Only**: All users may read records but owners and higher-level users may edit them. This is a typical use of this function whenever a set of data must be shared for collaboration but not everyone in the group must be able to edit it.

- **Public Read/Write**: All users can view and edit records. This is often used whenever data must be readily available and processed by many users.

In the following chapters, we will go into much more detail about how to configure OWDs and how you can apply them well in your organization, together with detailed examples for each setting.

Role Hierarchies

Role Hierarchies are another core part of Salesforce's sharing model. They mimic the reporting structure of an organization; data visibility automatically flows up the hierarchy. When a user is assigned to a role, all the records owned by users in roles beneath that role are accessible. The way this works ensures managerial and executive officials will be abreast of their subordinates' data, therefore making reporting easier and decision-making more efficient.

However, role hierarchies only show visibility of records; editing rights are not automatically granted there; that's where more advanced sharing mechanisms come in. Role hierarchies, done correctly, do eliminate the need for a manual intervention to give access at any level within an organizational structure.

For the rest of this book, we explain how to design and tune role hierarchies so that they meet the needs of your business.

Sharing Rules

While OWDs and role hierarchies form the foundation for what records people can access, Sharing Rules provide a way to extend access even further. Administrators can specify additional access based on owning records or qualifying against other criteria; an example of this could be a sharing rule that might provide a regional sales manager with access to opportunities owned by sales reps in different territories. There are two types of sharing rules:

- **Owner-Based Sharing Rules**: These rules would be sharing records based on who owns them. For instance, you can use this feature to share all the opportunities of a specific region to the other region's users.

- **Criteria-Based Sharing Rules**: These are particular rules in which sharing will be based on particular criteria of fields. For instance, one can create a sharing rule that shares access to all accounts marked as high priority regardless of who owns the account.

In Chapter 3, we will discuss the configuration of sharing rules to open up/widen this circle of people and teams with the appropriate level of access at the right time, with only the appropriate security controls.

Manual Sharing

Lastly, Manual Sharing provides an ability for a user to distribute individual records to other users or groups in an ad hoc manner. Once again, as an alternative to automated sharing options, such as OWDs, role hierarchies, or sharing rules, manual sharing is initiated by the record owner or an administrator who may wish to grant specific access to a particular record for selected users. This is useful when access is only needed for a single, exceptional case or even temporarily.

Manual sharing is usually applied in areas where automatic self-sharing mechanisms cannot meet the particular needs of a record. It also allows administrators or users to bypass normal sharing settings, so that data reaches the right people.

Why Do These Matter?

Bottom line, these four foundational elements of the Salesforce sharing architecture—OWDs, Role Hierarchies, Sharing Rules, and Manual Sharing—are key elements that work together to create a safe yet flexible model for controlling access to data. Together, they provide underpinnings about how an organization manages its data: right people getting the right information and out of the way of inappropriate users seeking sensitive data. While these components give you a pretty powerful framework, they are merely the foundation of a much more detailed approach tailored to securing your data.

As we go through the book, we will be discussing each one of these components in greater detail, but this time focusing on configuration techniques and best practices to ensure that your Salesforce environment is balanced with your business goals and secure.

Common Misconceptions About Data Security in Salesforce

Salesforce provides a very robust and multi-layered data security model, usable to give organizations highly customizable access control. Like all things that are flexible, there are also common misunderstandings that can create either security gaps or inefficiencies. In this chapter, we break down some of the most common myths associated with data protection in Salesforce by going into great detail on each one to better understand how to use Salesforce's tools for your benefit to achieve best-in-class and compliant data protection.

Myth 1: Single Setting Can Provide Full-Scale Data Protection

The most common myth about Salesforce security is that a single setting or permission can create a broad security blanket. But Salesforce takes its security on a layered basis and, hence, has different roles played by each layer. Making just one setting or permission set often leaves big holes open and vulnerable to sensitive data exposure.

Explanation

The three core layers in which data security operates within Salesforce include object-level, field-level, and record-level security. Every layer has a distinct role in protection, and the thorough deployment of all three layers at once ensures a solid framework.

Let's say we are working with a real estate company that tracks property listings, client contracts, and transactions details using Salesforce. At the object level, the company might decide that only agents and brokers should have access to the "Client Contracts" object, while other teams, such as marketing, are entirely restricted from accessing it. However, within the "Client Contracts" object, there could be specific fields, such as commission rates or contract values, that should only be visible to senior brokers, not junior agents. This is where field-level security comes into play.

On top of these restrictions, record-level sharing settings ensure that agents can only access client contracts related to their own deals, while managers might need visibility into all team transactions to oversee operations. For example, in certain cases, manual sharing may be utilized to permit finance personnel to access contracts that require review for a limited period while other records remain private.

As this example makes clear, reliance on any single layer alone—be it the object permissions layer, the field-level restrictions layer, or the record sharing settings layer— is likely to prove insufficient to meet the complexities of most business requirements. Each is designed to be a specific layer for a specific security requirement, and clarity about how these layers interrelate helps build an overall strategy about data protection. It is only when these layers can be combined that organizations can achieve a degree of flexibility and control that is necessary to protect sensitive information yet continue to allow workflow smoothness.

Myth 2: Sharing Rules Are Universally Valid

It's a misconception that the rules around sharing work as an all-things-to-all-people kind of thing, mainly imposing default permissions on everyone. This would expose sensitive information far too excessively or place it way out of reach for those who need it most. In reality, the engineering design of sharing rules is there to enhance basic security settings and provide specific, conditional access based on certain business needs.

Explanation

Sharing rules provide selective access to records beyond the constraints set by Organization-Wide Defaults and role hierarchies. They can operate like a controlled gateway and let one have access to selected data while keeping the integrity of the broader security framework intact.

For example, consider a retail company utilizing the application of Salesforce to guarantee its commodities are sold and to monitor client information. Sales staff can require such sensitivities as customer purchase histories—for example, how many products bought in a period or recurrence order frequency—in order to plan their moves properly. A marketing or logistics employee, who may not need such information, should not be allowed to access it.

In this particular scenario, rules regarding sharing could be applied to limit it only to the respective sales teams, while at the same time, deny access for everyone else. This means sensitive data would be protected, and at the same time, authorized users would still view that information to do their roles correctly.

Why Are Sharing Rules Not a "Set It and Forget It" Feature?

The other common misconception is that once shared, rules for sharing work independently. They do not work independently, but must instead intelligently align themselves with security configurations put in place before them, like OWDs and role hierarchies, to prevent accidental conflicts or overlaps.

For instance, consider the case where OWDs are set to "Private" for customers and a sharing rule is turned on for access by the sales team. The combination makes the balance of the framework in such a way that OWDs give baseline restrictions.

Sharing rules selectively expand access so that relevant users can perform their tasks.

This is possible by layering it, which will prevent over-sharing: the sensitive information should only be in the right hands, and by the same token, under-sharing, which will again negatively impact productivity when legitimate access is needlessly restricted.

Business Context

Sharing rules are best applied when the configuration is based on actual business needs. These should therefore be applied in very specific circumstances in order to address specific scenarios rather than be a blunt sword to an entire spectrum of problems.

For example, within a multinational company, sharing rules can be configured so that regional sales teams see data pertinent to their markets, but the other teams do not see some of the extra features which may not apply to those regions.

This specificity prevents possible data privacy risks while promoting cooperation in teams that truly require the data to carry out their functions.

Just the Right Balance

The right way to exploit sharing rules is within a balance that strikes

- **Flexibility**: To grant access for necessary purposes without lowering security measures

- **Precision**: To avoid over- or under-sharing

- **Alignment to Other Security Settings**: For a coherent design

This grants them room to build an evolutionary security model in line with the compliance model rather than taking the very safe approach of assuming the rules are universal and one-size-fits-all.

Myth 3: Field-Level Security Addresses All Data Protection Needs

Frequently viewed as a myth is that field-level security, on its own, will accomplish all necessary requirements to safeguard the information of an organization. Field-level security is, however, a powerful feature in that it enables administrators to define what can be viewed or updated as far as what fields within an object go, but this alone is only part of the total security solution.

Explanation

Record-level security is quite useful in concealing sensitive fields, like financial or medical data. But record-level security cannot protect your data if you just merely rely on record-level security. Let's take an example: a team of finance may need to gain access to most fields of a customer's account but fields containing his payment details are accessible only by a few selected roles in that team.

Knowing that field-level security needs to be combined with other Salesforce settings, administrators are saved from situations that could inadvertently expose sensitive information.

For example, consider a government agency using Salesforce to manage records of its citizens. While field-level security may restrict users from viewing fields like "social security number" or "income," it may be that users are still able to view a record containing these fields but would not be able to see the following fields. This layered approach simultaneously governs both the visibility of particular data points and access to complete records.

Myth 4: Manual Sharing Offers Unlimited Access to Records

Manual sharing is very convenient and simple without causing any effects on the overall organization-wide settings. In this regard, it is thought that in manual sharing, the same records are accessed without any restrictions. Not so. This is not true.

Explanation

Manual sharing can only provide record-level access. That is, it allows a user to read or edit a record but doesn't extend any further to any permission at an object or field level beyond that.

For example, if a user is not allowed to edit an object through his profile or permission set, no manual sharing of a record inside that object gives him the right to edit. They can see the record but still respect those actions that settings on their profile prohibit them from doing.

So, let's say you manage donor information for a nonprofit in Salesforce, and one of your organization's fundraisers needs temporary access to view the record of one particular donor. That means an administrator needs to manually share that access. Since the profile of that member does not have the authority level to edit the field called "Donation Amount," that member, given only the access for that particular record, will not be able to do so. So, no need to worry: access sharing is guaranteed to take place properly, without permission bypass at profile levels or even field-level basis.

Myth 5: Profiles and Permission Sets Secure All Data Completely

At the center of the architecture of permission in Salesforce is actually profiles and permission sets; because of this, people often use them as a full solution. They really enforce at the object and field levels rather than at a record level. Profiles and permissions sets aren't really actually locking down access to data.

Explanation

Although it is actually true that profiles and permission sets control, from an object point of view, whether or not a user can read, create, edit, or delete information, the record-level access is controlled by permitting another configuration, such as role hierarchies, sharing rules, and manual sharing that determines record-level permissions of which records, within an object, a user is really entitled to view. So, profiles and permission sets alone would likely lead either to accidental loss of access or overexposure.

For example, then, a national sales manager and a local sales representative might need to access the same object, "Opportunities," but the national manager needs to see all opportunity records, whereas the representative only needs to view records relevant to his or her region. Visibility should not be over-extended or unduly restricted if access controls are strictly based on profiles or permission sets, which is why role hierarchies and sharing rules have to be layered onto those basic configurations.

Once the purpose of profiles and permission sets is understood, administrators are less prone to configuration mistakes and can instead design a comprehensive access structure that includes features drawn from multiple security settings to achieve optimal and more robust data protection.

Myth 6: Role Hierarchies Define Data Access Only

Role hierarchies are often mistaken as only a definition of data access structure, especially in highly structured reporting scenarios. But role hierarchies define only what can be viewed with regard to access—not what can be done on records.

Explanation

Role hierarchies allow users in a higher role to see the data of the other users who have a lower role. This can be very useful where visibility needs to mirror the hierarchical management, for example, if you need to monitor customer interactions coming from the people reporting to the manager. Meanwhile, while role hierarchies do allow visibility, they inherently don't assign edit rights. The edit rights must still be configured in roles or permission sets to actually do something with the record.

For example, take a global company with regional managers and sales forces who operate on a global scale. Through the help of role hierarchy, the regional manager can view all records owned by his or her sales team's members; however, he or she would need the object permission in that regard to delete or edit them. Otherwise, role hierarchy would only let view the record but never interact more than their profile permits.

Myth 7: Organization-Wide Defaults (OWDs) Are Enough for Data Security

Most people believe that by implementing the OWDs, they are setting up proper data security configuration. That is utterly not true. OWDs only take into account the base level of access at which the whole organization is set. These are merely foundations of the security model; much has to be configured in order for them to be fully available.

Explanation

OWDs set a default level of access for every object throughout an entire organization and can make records Public Read/Write, Public Read Only, or Private. However, OWDs are not a security measure unto themselves. They are instead one piece that, when used with role hierarchies, sharing rules, and manual sharing settings, defines specific pathways of access within an organization.

For example, think of a financial services company that sets the OWD of a client file to Private. Now, client data will be accessible by default only to certain individuals but must often be shared with customer service representatives or compliance officers. That's the beauty of layering OWDs over other types of access settings: it enables an organization to build a more complete and secure data access model.

Myth 8: Audit Logs Are Enough for Security Monitoring

Almost every organization deludes itself at the end by thinking that Salesforce will take care of security through its in-built audit logs. Audit logs can be used to track who accessed or changed what, but they alone will not prevent security risk nor ensure proactive monitoring. Although this is very useful to know who accessed or made what changes in the data, audit logs prevent no security risks and cannot be used proactively to monitor security issues.

Explanation

The Salesforce system, as data and configurations capture historical changes, also becomes a good source for the analysis of incidents—audit logs. However, because audit logs merely record, it cannot prevent someone who does not wish to go beyond access settings from accessing data or configurations beyond his intents due to lack of proactive monitoring or reviews by the user on access settings. There has to be regular checks by access configurations as business needs change through time.

For instance, an organization may check role hierarchies and sharing settings every quarter to detect any unwanted permissions that have accidentally been granted over time. In short, it implies not only knowing what has happened in the past through audit logs but also having an active stance on data security.

Myth 9: Field-Level Security Not Required If Object Access Is Already Controlled

One more myth is the statement of not considering field-level security if users do not have object access. But it is critical even when there is proper assignment of object-level permissions. Some integration/reports will also accidentally leak field-level information, and therefore, securing sensitive fields, like financial information or personal detail does matter, mostly for purposes of remaining well within acceptable regulatory limits.

Explanation

Object-level security determines whether an end user can view, create, edit, or delete records for the object as a whole. However, merely securing access to an object does not limit users from viewing sensitive fields in those records. Nor are users who do not directly access an object through the Salesforce interface necessarily prevented from indirect access to its sensitive fields through

- **Reports**: When a report is generated that fetches information from an object, users will be able to see fields they shouldn't.

- **Integrations**: Any external systems that are integrated with Salesforce have the potential to fetch sensitive data because they work by bypassing the UI controls.

- **APIs**: Any application working with Salesforce through APIs will allow sensitive fields to be fetched, if their configuration isn't appropriate.

Field-level security ensures that sensitive data remains secure even in indirect access scenarios.

Myth 10: Security Settings Are Only Set Once, and There Is No Need to Review Them Further

One of the most pervasive myths about Salesforce security is the mentality that once data security and sharing settings are set, the organization no longer needs to pay any more attention to them, which often means a set and forget" mentality. Although these settings do provide an excellent foundation to control the access, the approach they translate to is static, whereas actual business operations, regulatory requirements, and growth in an organization can all become dynamic.

Without scheduled reviews and updates, security configurations become stale, leading to unintentional over-exposure or underutilization of data. This myth is fraught with danger since teams are rapidly growing, roles are changing daily, and regulations keep getting heavier.

Explanation

Data security in Salesforce isn't a one-off exercise—it is periodic. The following factors enlist why it is necessary for doing periodic reviews:

Business needs change with time. Organizations expand and restructure or alter their workflows. Security configurations that suited a smaller group or simplified structure might no longer align with current operations.

- **Regulatory Updates**: GDPR, HIPAA, CCPA are laws that frequently change. Hence, the compliance needs are always evolving, and one might have to make changes based on these regulations for sharing and visibility.

- **Role Changes**: Access requirements or restrictions may change as and when roles and responsibilities change within teams.

- **Inclusion Growth**: New integrations of external tools that interact with Salesforce might require adjustment in making sure data exchange is securely made without revealing any sensitive information.

Brainstorming

1. **What is the primary purpose of Salesforce's multi-layered data security model?**

    A) To prevent data duplication

    B) To ensure scalability across Salesforce orgs

    C) To allow precise control over data visibility and access

    D) To facilitate data backups efficiently

    Answer: C

2. **Which of the following is NOT a layer in Salesforce's data security framework?**

    A) Object-Level Security

    B) Field-Level Security

    C) Record-Level Security

    D) Region-Level Security

    Answer: D

3. **What governs whether a user can access specific fields within an object?**

    A) Role Hierarchies

    B) Field-Level Security

    C) Organization-Wide Defaults

    D) Sharing Rules

    Answer: B

4. **Which security layer ensures that only certain records of an object are visible to a user?**

    A) Object-Level Security

    B) Record-Level Security

    C) Sharing Rules

    D) Profiles

    Answer: B

5. **What is the default data-sharing mechanism in Salesforce that determines baseline access?**

    A) Organization-Wide Defaults (OWDs)

    B) Role Hierarchies

    C) Field-Level Security

    D) Permission Sets

    Answer: A

6. **True or False: Field-level security applies even if a user has access to the object.**

    A) True

    B) False

    Answer: A

7. **A sensitive data field is appearing in reports and downstream integrations for users who should not see it. Which security setting is the most likely root cause?**

   A) Object-Level Permissions

   B) Field-Level Security

   C) Sharing Rules

   D) Manual Sharing

   Answer: B

8. **When records are shared to a public group using a sharing rule, what determines whether users get Read-Only or Read/Write access?**

   A) The public group type

   B) The sharing rule's access level setting

   C) The user's profile

   D) The role hierarchy

   Answer: B

9. **Why is it essential to regularly review sharing and visibility settings in Salesforce?**

   A) To ensure system performance is optimal

   B) To update permissions based on organizational changes and evolving regulations

   C) To align Salesforce features with marketing campaigns

   D) To avoid data duplication

   Answer: B

10. **Which Salesforce feature allows sharing settings to be extended beyond OWDs and role hierarchies?**

    A) Permission Sets

    B) Sharing Rules

C) Field-Level Security

D) Manual Sharing

Answer: B

11. **What is a common misconception about Salesforce sharing rules?**

    A) They only apply to external users.

    B) They override Organization-Wide Defaults.

    C) They provide one-size-fits-all access.

    D) They do not require configuration.

    Answer: C

12. **What governs object-level access in Salesforce?**

    A) Profiles and Permission Sets

    B) Role Hierarchies

    C) Sharing Rules

    D) Field-Level Security

    Answer: A

13. **Which scenario demonstrates a need for role hierarchy?**

    A) A manager needs access to all records their team owns.

    B) A compliance officer requires audit logs across all objects.

    C) A marketing user requires specific campaign data.

    D) A vendor requires access to service contracts only.

    Answer: A

14. **True or False: Sharing rules can decrease access levels granted by OWDs.**

    A) True

    B) False

    Answer: B

15. What's the risk of not considering field-level security when object permissions are granted?

A) Users cannot view the object

B) Sensitive fields might be unintentionally exposed

C) Reports won't function properly

D) Integration with external systems fails

Answer: B

16. How do Organization-Wide Defaults (OWDs) affect record access?

A) They define default access levels for all users across records of an object.

B) They provide granular control over field access.

C) They restrict access to specific objects only.

D) They allow users to edit their personal data only.

Answer: A

17. What is the benefit of using a role hierarchy in Salesforce?

A) Reduces the number of profiles needed

B) Automatically grants access to records owned by subordinates

C) Enhances field-level security

D) Simplifies object-level permission configuration

Answer: B

18. True or False: Field-level security settings are ignored for API integrations.

A) True

B) False

Answer: B

19. **Which of the following is an example of a regulatory requirement that influences data-sharing decisions?**

    A) HIPAA

    B) GDPR

    C) CCPA

    D) All of the above

    Answer: D

20. **How can manual sharing be used effectively in Salesforce?**

    A) To automate access for all users

    B) To grant temporary access to specific users on a case-by-case basis

    C) To override field-level security

    D) To replace sharing rules entirely

    Answer: B

21. **Why is it important to identify stakeholders during data-sharing decisions?**

    A) To ensure stakeholders are involved in system audits

    B) To align data-sharing settings with business needs and minimize risks

    C) To reduce the administrative workload of IT teams

    D) To manage data duplication across objects

    Answer: B

22. **True or False: Profiles and permission sets alone ensure complete data security.**

    A) True

    B) False

    Answer: B

23.   What's a common myth about Salesforce data security?

A) Object-level security can override record-level settings.

B) Role hierarchies automatically grant too much access.

C) Security settings are "set and forget."

D) Sharing rules eliminate the need for field-level security.

Answer: C

24.   What happens if OWD is set to "Private" for an object?

A) All users can see all records of that object.

B) Only the record owner and users above them in the role hierarchy can access the records.

C) All users can edit but not delete the records.

D) The object becomes inaccessible to all users.

Answer: B

25.   Which tool is best for creating fine-tuned, temporary access to records?

A) Permission Sets

B) Field-Level Security

C) Manual Sharing

D) Sharing Rules

Answer: C

26.   What is the relationship between role hierarchies and record-level security?

A) Role hierarchies override field-level security.

B) Role hierarchies determine access to records owned by subordinates.

C) Role hierarchies are the only way to grant access to external stakeholders.

D) Role hierarchies apply only to objects with OWD set to "Public Read/Write."

Answer: B

27. **Which of the following is NOT impacted by sharing rules?**

A) Read/write access to records

B) Visibility of fields within a record

C) Specific access for public groups or roles

D) Permissions that supplement OWDs

Answer: B

28. **True or False: Permission sets can be assigned to multiple users at once.**

A) True

B) False

Answer: A

29. **What is one primary reason to review and update sharing settings regularly?**

A) To improve Salesforce system performance

B) To comply with changing regulations and business needs

C) To eliminate unnecessary objects

D) To merge duplicate records

Answer: B

30. **How does Salesforce help manage data sharing across regions with differing regulations?**

A) By providing customizable sharing settings at multiple levels

B) By enforcing a single global sharing rule

C) By automatically detecting regulatory violations

D) By limiting data access to administrators only

Answer: A

31. **What is the key purpose of field-level security in Salesforce?**

   A) To hide objects from unauthorized users

   B) To prevent accidental deletion of records

   C) To control access to specific fields within a record

   D) To restrict API access entirely

   Answer: C

32. **Which feature ensures that only appropriate users can view a record in a "Private" OWD setting?**

   A) Permission Sets

   B) Role Hierarchies

   C) Field-Level Security

   D) Public Groups

   Answer: B

33. **What is a drawback of relying only on OWD for security settings?**

   A) It grants overly broad access to fields.

   B) It does not account for granular field-level needs.

   C) It cannot differentiate between internal and external users.

   D) It is incompatible with sharing rules.

   Answer: B

34. **What is a best practice when configuring sharing rules?**

   A) Use sharing rules as a replacement for OWD settings

   B) Create rules based on granular business needs and stakeholder roles

   C) Apply sharing rules to all objects without exception

   D) Disable role hierarchies to streamline rule application

   Answer: B

35. **True or False: A user can access fields hidden by field-level security through reports.**

    A) True

    B) False

    Answer: B

36. **Why is it important to map user roles to data needs?**

    A) To ensure that users can access all data in the org

    B) To prevent duplication of records across users

    C) To align permissions with business functions and minimize risk

    D) To automate the setup of sharing rules

    Answer: C

37. **What should be considered when granting access to external stakeholders like vendors?**

    A) They must be assigned full access to avoid delays.

    B) Their access should be limited to specific data relevant to their role.

    C) They should always be part of the internal role hierarchy.

    D) No additional configurations are required for external stakeholders.

    Answer: B

38. **What is the role of public groups in Salesforce sharing settings?**

    A) To manage access across large user groups efficiently

    B) To enforce default sharing rules

    C) To override field-level security

    D) To grant full administrative rights

    Answer: A

39.   Which of these is NOT a valid use of field-level security?

A) Hiding social security numbers from non-administrative users

B) Restricting API access to specific fields

C) Controlling the visibility of records within a public group

D) Limiting access to financial details to only a specific team

Answer: C

40.   What does "least privilege" mean in Salesforce data security?

A) Assigning as few permissions as possible to each user

B) Granting only the minimum access needed for a user to perform their role

C) Restricting all users to read-only access

D) Automating permissions based on user location

Answer: B

41.   What's a key risk of using outdated sharing and visibility settings?

A) Overexposure or underexposure of critical data

B) Increased duplicate records

C) System performance degradation

D) Reduced role hierarchy efficiency

Answer: A

42.   Which of these is an example of improper data-sharing configuration?

A) A public group is created for team collaboration.

B) Sensitive financial fields are visible to all users due to inadequate field-level security.

C) Role hierarchy grants a manager access to their team's records.

D) Sharing rules supplement OWDs to meet business needs.

Answer: B

43. **True or False: Field-level security is sufficient on its own to secure Salesforce data.**

    A) True

    B) False

    Answer: B

44. **Which of these describes a common misconception about Salesforce data security?**

    A) Sharing rules can only extend access, not restrict it.

    B) Role hierarchies are only useful for administrators.

    C) Profiles and permission sets are sufficient for complete security.

    D) Field-level security settings apply to reports.

    Answer: C

45. **What is the relationship between field-level security and object permissions?**

    A) Field-level security applies only if object access is denied.

    B) Field-level security works in conjunction with object permissions to control access.

    C) Object permissions override field-level security.

    D) Object permissions do not impact field-level settings.

    Answer: B

46. **Which of the following is true about record-level security in Salesforce?**

    A) It only applies to objects with OWD set to Public Read/Write.

    B) It governs which individual records a user can access within an object.

    C) It overrides all field-level security settings.

    D) It is configured automatically when an object is created.

    Answer: B

47. **What is the purpose of manual sharing in Salesforce?**

A) To automatically grant access to all records in an object

B) To provide record access for specific users when no other sharing rule applies

C) To restrict access to sensitive data for a group of users

D) To enforce field-level security settings across multiple users

Answer: B

48. **When should sharing rules be used in Salesforce?**

A) To replace role hierarchies for managing access

B) To grant access to users based on record ownership or criteria

C) To restrict access beyond the OWD settings

D) To assign object-level permissions to users

Answer: B

49. **What does OWD stand for in Salesforce security settings?**

A) Organization-Wide Data

B) Organization-Wide Defaults

C) Object-Wide Defaults

D) Object-Workflow Design

Answer: B

50. **Why is it important to regularly audit sharing and visibility settings?**

A) To improve system performance and reduce API usage

B) To ensure compliance with evolving business needs and regulations

C) To consolidate profiles and permission sets

D) To merge duplicate records

Answer: B

51. **What happens if a user has no access to a field through field-level security?**

A) They can still see the field in reports.

B) The field is hidden from view in both the UI and API.

C) They can edit the field but not view its data.

D) They are automatically assigned access to the field.

Answer: B

52. **True or False: Role hierarchies can restrict access to records for users higher in the hierarchy.**

A) True

B) False

Answer: B

53. **Which Salesforce feature allows users to define access based on geography or department?**

A) Role Hierarchies

B) Territory Management

C) OWD Settings

D) Field-Level Security

Answer: B

54. **What's the difference between profiles and permission sets?**

A) Profiles are assigned to roles, while permission sets are assigned to objects.

B) Profiles define baseline access, while permission sets extend or add permissions.

C) Profiles manage field-level security, while permission sets manage OWDs.

D) Profiles are for external users, and permission sets are for internal users.

Answer: B

55. **Which of the following is NOT true about sharing rules?**

    A) They can only extend access, not restrict it.

    B) They supplement role hierarchies and OWDs.

    C) They are applied automatically to all objects.

    D) They can grant access based on roles, public groups, or territories.

    Answer: C

56. **What's a key risk of granting "View All" or "Modify All" permissions at the profile level?**

    A) It may reduce system performance.

    B) It could lead to excessive access to sensitive data.

    C) It overrides sharing rules, but not field-level security.

    D) It automatically disables manual sharing options.

    Answer: B

57. **What is a common misconception about manual sharing?**

    A) It can override OWDs and role hierarchies.

    B) It applies globally to all users.

    C) It is a long-term solution for access management.

    D) It is only available to administrators.

    Answer: C

58. **In Salesforce, which setting should you use to ensure a specific user group has read-only access to certain records?**

    A) Field-Level Security

    B) Sharing Rules

    C) OWD set to Public Read/Write

    D) Manual Sharing

    Answer: B

59. **What is the impact of disabling the role hierarchy in a sharing setting?**

    A) It prevents users higher in the hierarchy from automatically accessing records.

    B) It disables OWD settings for that object.

    C) It limits access to only public groups.

    D) It removes all permissions for users in that role.

    Answer: A

60. **What should be done when new business regulations impact data-sharing requirements?**

    A) Increase the number of profiles and permission sets

    B) Regularly review and update sharing and visibility settings

    C) Set OWD to Public Read/Write for simplicity

    D) Use manual sharing as the primary configuration

    Answer: B

CHAPTER 2

Profiles and Permission Sets: Unlocking the Power of Access Control

Access control is at the core of every Salesforce implementation, because that's what allows the right users to have the right tools and the right data to do their jobs in a better way while keeping sensitive information safe. Without proper access control, organizations are exposed to operational inefficiencies, data breaches, and compliance failures. This chapter will lay the groundwork for digging deep into some of the most fundamental Salesforce access control tools: Profiles and Permission Sets.

The underpinnings on which Salesforce builds its robust security model involves Profiles and Permission Sets. These are tools used to help define, manage, and fine-tune user permissions. While Profiles establish the base-level permissions for a user, Permission Sets provide for even more layers of permissions, so that there is unique, adaptive control over what can be done and seen in Salesforce. This will enable you to design a system that is both secure yet efficient enough to dynamically adapt to the ever-changing needs of your organization—by understanding how these tools function individually and in concert with each other.

Why Profiles and Permission Sets Matter

Access control in Salesforce is far from a nice-to-have; it is essentially an important component of strong systems governance. Poorly managed permissions open up security vulnerabilities, enable data misuse, and sow operational chaos. Conversely, access control is realized when the power to influence and control information and the systems that provide it is made available to only those employees who need it to do their work.

© Sandhya Sharma 2026

S. Sharma, *Unlocking Salesforce Data*, https://doi.org/10.1007/979-8-8688-2305-3_2

- **Data Integrity Is Protected**: Sensitive information is protected from unauthorized access, reducing risk and building trust.

- **Compliance Is Supported**: Organizations can support regulatory as well as internal standards by limiting who can access certain data and perform certain actions.

Profiles and Permission Sets enable administrators to balance these competing priorities in a structured yet flexible way to define who can access what.

Roles of Profiles and Permission Sets

At first glance, Profiles and Permission Sets seem to serve roughly the same purpose. Both enforce access to data and functionalities within Salesforce. However, their differences make them so effective when used together.

Profiles are the foundation of any access control implementation, defining a user's default permissions. A Profile determines what objects, fields, and applications a user can work with and what a user can do to those objects. Every Salesforce user must be assigned a profile, and this assignment provides a consistent baseline for permissions.

Permission Sets extend the permissions granted by profiles; these permission sets add additional permissions and never remove anything. This is very important to manage exceptions, sometimes temporary needs, or truly specialist roles requiring more permissions than their profile.

For instance, a Field Technician profile might give her access to all objects such as Work Orders and Service Appointments that are at the heart of her job: managing tasks on-site and scheduling. However, if an engineer temporarily takes charge of managing inventory for a particular project—for instance, tracking spare parts and updating the stock levels—a Permission Set may be utilized to provide them with extra access to features for Inventory Management, as that would be required for the project. That way, they would be provided with the requisite tools without changing the basic Profile shared by all engineers, while also ensuring system consistency, and addressing this specific requirement.

Profiles vs. Permission Sets: Interplay

The interplay between Profiles and Permission Sets is where the power of Salesforce's access control system really rests, as it allows administrators to implement a multi-layered and dynamic permission management approach.

For example, a Profile might be the starting point, giving a user access to a specific role's core functionalities. Permission Sets can then be layered on top to address those unique scenarios, such as granting cross-functional teams' access to shared objects during a project or providing temporary permissions for a short-term initiative.

This layered approach also helps ensure that permissions are both consistent and flexible. The users obtain the access needed without over-provisioning, thereby ensuring that the risk of unauthorized action is limited while maintaining system efficiency.

A good access control system should be robust: it balances security, usability, and scalability. With the growth and changes that accompany organizations, environments in Salesforce typically get more and more complex with respect to the new users and roles, and business requirements emerge regularly. In response to this, Profile and Permission Set management must be proactive.

This chapter provides a roadmap to the implementation of these principles, guiding you through developing an access control strategy that is at once robust and flexible.

Conclusion

Profiles and Permission Sets are the basis for the security model of Salesforce. Profiles and permission sets form a firm combination of structure along with flexibility, thus helping you create an access control mechanism that would align with your organizational objectives as well as accommodate change while keeping data secure.

In this chapter's introduction, we lay the groundwork for our detailed exploration to come. In the following sections, we will go deeper into how to configure Profiles and Permission Sets, best practices for their application, and strategies for a scalable and efficient system. Whether you're an experienced administrator or new to Salesforce, this chapter will give you the know-how and tools to make the controls on access your own.

Understanding Profiles and Permission Sets

Users need profiles and permission sets as the foundation to set, manage, and fine-tune user permissions. Profiles set the basic levels of access while permission sets allow for adding additional permissions on an ad hoc basis for each user. Together, they provide this layered and flexible structure that balances organizational security with user productivity. In the following sections, we'll take a closer look at how they work, interact, and what best practices to follow when applying a good strategy for access control in the following section.

Key Characteristics of Profiles

Each user in Salesforce needs to be assigned at least one Profile, and that will be their baseline of permissions and access to the system. A Profile defines the default capabilities of the user: which objects, fields, applications, and features may be accessed and what actions may be performed.

Baseline Permissions

Profiles form the first level of access control, as they define what is available, including access granted at the object and field levels. For instance, a "Sales Representative" Profile would likely be allowed to access Leads, Opportunities, and Accounts, depending on the specific permissions: Create, Read, Edit, or Delete.

Mandatory Assignment

Every Salesforce user must have exactly one Profile, which means it's an integral part of the access control system, ensuring that no user operates without defined permissions. As it can be seen in the Figure 2-1 below, a Profile is mandatory for user creation.

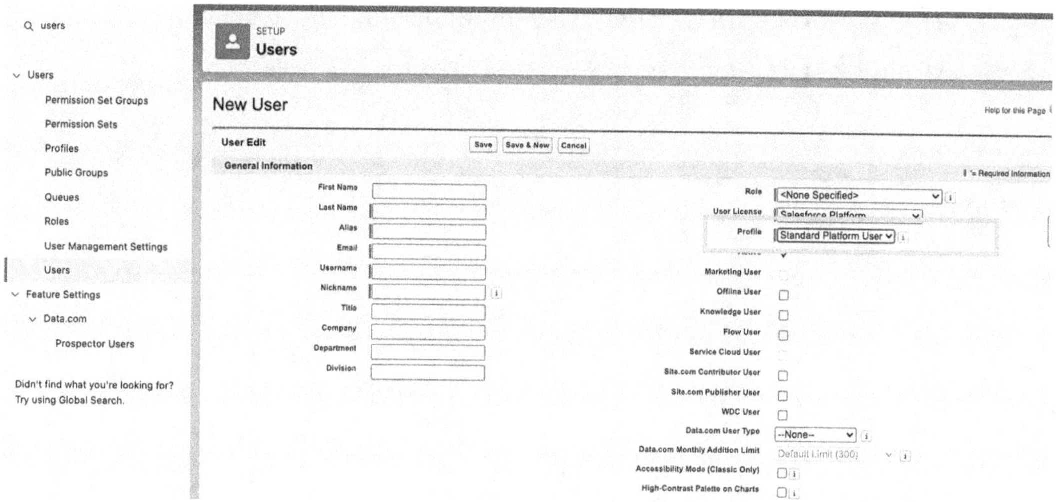

Figure 2-1. *User creation UI in Salesforce org*

Scope of Permissions

Profiles have numerous permissions, such as the following:

- **Object Permissions**: Control over standard and custom objects (Figure 2-2)

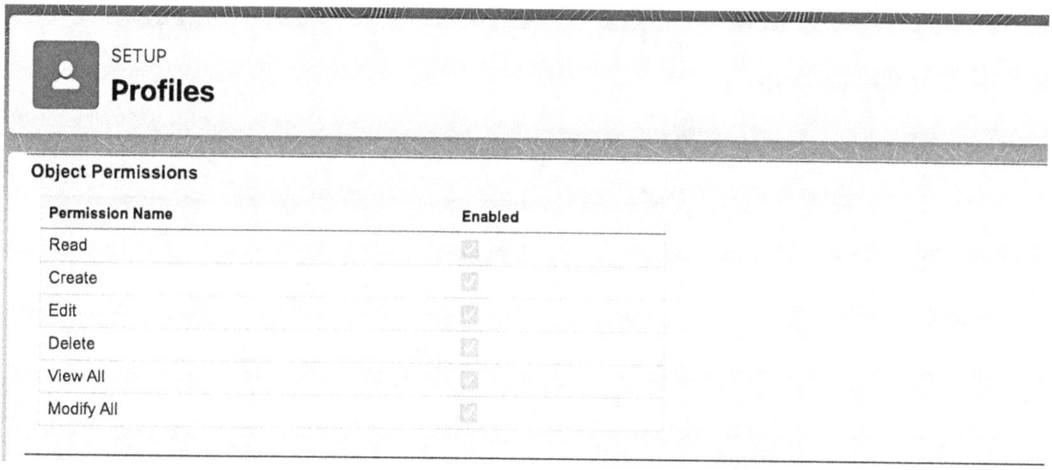

Figure 2-2. *Object-level permission user interface in profile*

- **Field-Level Security**: Defines visibility and editability of fields (Figure 2-3)

Field Permissions

Field Name	Field API Name	Read Access	Edit Access
Account Name	Name		
Account Number	AccountNumber		
Account Owner	OwnerId		
Account Record Type	RecordTypeId		
Account Site	Site		
Account Source	AccountSource		
Active	Active__c		
Annual Revenue	AnnualRevenue		
Billing Address	BillingAddress		
Channel Program Level Name	ChannelProgramLevelName		
Channel Program Name	ChannelProgramName		
Clean Status	CleanStatus		
Created By	CreatedById		
Customer Portal Account	IsCustomerPortal		
Customer Priority	CustomerPriority__c		
D&B Company	DandbCompanyId		
Data.com Key	Jigsaw		
Description	Description		
D-U-N-S Number	DunsNumber		
Einstein Account Tier	Tier		
Employees	NumberOfEmployees		

Figure 2-3. *Field-level permission user interface in profile*

- **Tab Settings**: Determines whether a tab is Default On, Default Off, or Hidden (Figure 2-4)

Profile
System Administrator

Q Find Settings   ✖ | Clone   Edit Properties

Profile Overview > Object Settings ▼ Accounts ▼

Accounts   Edit

Tab Settings
Default On

Figure 2-4. *Tab setting user interface in profile*

- **System Permissions**: Controls abilities like "Export Reports" or "View All Data" (Figure 2-5)

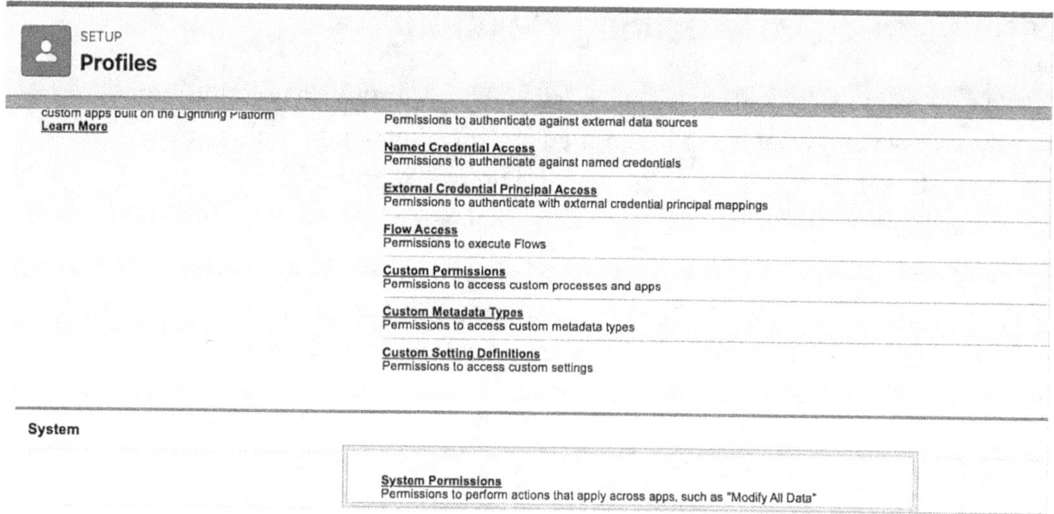

Figure 2-5. *Profile overview user interface*

Standard vs. Custom Profiles

Salesforce comes with a number of pre-built Standard Profiles: System Administrator, Standard User, and Marketing User. While they are useful starting points, in most business-specific cases, they don't have the degree of granularity needed.

But with Custom Profiles, administrators can create unique roles within their organizations that may include specific permissions tailored to the needs of the organization. For example, you could have a "Regional Sales Manager" Profile that differs from that of a "Sales Representative."

Restrictions of Profiles

Profiles are very essential but not without their limitations:

- They do not lend themselves well to exceptions or to temporary permissions.

- Changing a Profile impacts each user assigned to it, which can cause problems to occur.

- Profiles are rather better for high-level, role-based permissions rather than detailed access control.

Permission Sets: Providing Flexibility

Permission Sets add to the functionality of Profiles by granting a user additional permissions beyond what his or her particular profile allows. It is additive in nature—it cannot take away a user's rights but can add them. Salesforce has introduced Permission Set Groups to make it easier to manage multiple Permission Sets. Through this, administrators can package the Permission Sets related to each other in one group and reduce administrative overhead. For example, a "Field Service Team" group might include permission sets for mobile app access, territory management, and work order approvals.

Main Features of Permission Sets

Each of the following features describe how permission sets enable an administrator to have control over access control in Salesforce and provide flexibility and precision without downgrading security or operational efficiency.

Granular Control

Permission Sets enable administrators to provide specific permissions to a particular user. For example, you might use a Permission Set to extend a Sales Representative's access to a custom report, but then leave the rest of the profile unchanged.

Multiple Assignments

While a single profile serves each user, users can be assigned more than one permission set. This allows for flexibility in meeting unique or changing needs without adjusting the profile.

Access for a Limited Time

Permission Sets are a very efficient way of providing temporary access for project or time-bound tasks. For example, suppose a project manager would need access to a custom reporting dashboard for the duration of three months to complete a project. A Permission Set can then be assigned that gives access to this custom reporting dashboard. When the project is done, the Permission Set can be revoked and permissions are restored promptly without creating unauthorized baseline Profiles for the manager.

Customized or Special Roles for a Specific Number of Users

Permission Sets are best used to provide for specialized roles that cannot easily be accommodated in Profiles. Consider an application where there is, for example, a small number of engineers who need to access advanced diagnostic tools within Salesforce. Instead of modifying the Profile shared by all engineers, a Permission Set can be assigned to this subset, enabling their specialized functions while maintaining consistency for others.

Permission Set Groups

Salesforce has Permission Set Groups, which basically make it easier to manage complex permission structures by grouping related Permission Sets together. A "Marketing Campaign Team" Permission Set Group may pool several related permissions to access Campaigns, Marketing Cloud features, and custom analytics dashboards into one location. That saves administrative effort and keeps the permissions of users with similar roles consistent and free of error, making it more efficient.

Profiles vs. Permission Sets

While Profiles and Permission Sets serve similar purposes, they play distinct roles in Salesforce's access control framework as outlined in Table 2-1.

Table 2-1. *Profiles vs. Permission Sets: Defining Baseline Access and Extending User Permissions in Salesforce*

Aspect	Profiles	Permission Sets
Purpose	Define baseline permissions for a user.	Grant additional permissions beyond the baseline.
Assignment	Every user must have one Profile. And a user cannot have more than one profile assigned.	Users can have multiple Permission Sets.
Scalability	Less scalable for handling exceptions	Scalable for addressing unique scenarios.
Login IP Ranges	Can be configured to restrict user logins to specific IP ranges.	Not available; cannot configure login IP restrictions.

Practical Use Cases: Coordinating Access with Profiles and Permission Sets

The following cases allow one to easily extend permissions while keeping the core Profile highly consistent within the organization

Use Case 1: Managing Cross-Functional Projects

A company starts a new marketing campaign that needs the cooperation of the Sales, Marketing, and Customer Support teams. Each department maintains its own Profiles, which are specific to each team's standard roles, but Permission Sets are used to provide shared access to a custom "Campaign Dashboard" during the project cycle.

Use Case 2: Temporary Access for Contractors

A contractor is engaged to run the company's sales pipeline through an audit. Instead of creating a new profile, they are given a permission set, which grants them temporary access to reports and dashboards. That way, they can perform their tasks without over-provisioning.

Use Case 3: Advanced Permissions for Managers

A Customer Success Manager might occasionally require access to a Knowledge Base, in order to review or update articles for their team. While their Profile offers Cases and Accounts to enable them to manage customer relationships, a Permission Set could be used to give them access to Knowledge Management features, so the manager could contribute to improving the Knowledge Base without affecting the default permissions applied to other managers in similar roles.

Emerging Trends

Salesforce continuously evolves its access control feature by introducing tools such as Salesforce Optimizer and AI-based insights. Such tools can introduce the need for easier detection of optimum configurations, pointing out security vulnerabilities easily, and permission management becomes seamless. Profiles and Permission Sets are the backbone of access control in Salesforce, offering structure and flexibility necessary to manage myriad user requirements. Once you understand roles and functionality, combined with best practices, you will build a secure, scalable system capable of keeping pace with organizational growth and changing needs.

Object-Level Security: Defining Data Access at the Core

Object-level security is the first layer of Salesforce's security model. It defines a user's access to specific objects—such as Accounts, Contacts, or Opportunities—determining whether they can Create, Read, Update, or Delete (CRUD) records within each object. This level of security is foundational because it provides the primary gatekeeping for access, preventing unauthorized users from interacting with certain types of data.

CRUD Permissions

Profiles and Permission Sets in Salesforce allow most of the Create, Read, Update, and Delete (CRUD) permissions. Each permission means a certain level of access to the Salesforce object and form the core of controlling the access to object level.

- **Create**: A user with Create access can create new records on an object. So, if a user has Create access to the Account object, then that user can add new customer accounts to Salesforce but perhaps cannot see other people's accounts depending on what record-level access was granted to that user.

- **Read**: A user can read records in an object and then make use of the same object to view the records without altering them. For example, if a user has a Contact object in Read-only permissions, the user can see information about contacts, but they can't delete, add any, or edit any contact on their own.

- **Update**: Update permission allows users to edit records. For example, a user with update permissions on the Case object can modify the details of existing cases, for instance, update the status of a case or add internal notes. If users don't have update permission, they can't make any changes to those records which they can view.

- **Delete**: This permission enables the deletion of records. Generally, this is not granted because accidental loss of information can result from record deletions, especially on objects with business-critical data such as Accounts and Opportunities. However, only users having Delete access, including administrators, can fully delete records.

At the most basic level, Profiles control the CRUD permissions, defining standard access levels a user should have based on their roles, being for example, a Sales Representative, Customer Support Agent, or a System Administrator. More granular permission assignment can be provided because permission exists besides that of profile-based where the administrator can still assign extra permissions above and beyond profile-based permissions and without having to set up multiple profiles.

Object-Level Security Requirements in Real-Life Scenarios

Object-level security in Salesforce determines what users can do with records of a particular object—such as creating, reading, editing, or deleting them. In real-world implementations, these decisions are driven by business roles, data sensitivity, and compliance needs, making it essential to design access thoughtfully before configuring profiles or permission sets.

Financial Data of Regulatory Industries

Strict regulations, for instance, HIPAA in financial services and healthcare, or GDPR, require careful control over access to sensitive information. For example, a bank might allow its customer service representatives to have read-only rights on the customer's financial accounts and provide account managers with the rights to create and update. Deletion permission probably could be limited to some small group of administrators in order not to let inadvertently deleted data make its way into backup storage.

An officer responsible for compliance may need to view the customer's account information but should not have the authority to update or delete it for auditing.

Read permissions only should be afforded to the compliance officer's Permission Set, which, therefore, complies with data protection laws.

Field-Level Support Team

An implementation of the support request handling features may involve a customer service team responsible for technical issues needing only Read and Update permissions on the Case object for case management without adding or removing any cases.

Support agents with Read and Update permissions keep track of and solve customer cases. If an agent mistakenly tries to delete the case, they will not succeed based on permission settings that guard the continuity of support data.

Field-Level Security in Salesforce: Fine Grain on the Precision of Access Control

Fine-grain access control is enabled in administration using field-level security in Salesforce. Administrators can control the fine-grain data access because, through the use of field-level security, they can manage access at the field level within objects.

Though object-level permissions grant general permissions to entire objects, field-level security provides a more granular layer of specificity wherein the level of access can be granted to particular fields or visibility or editing restrictions put in place based on user roles or profiles. This approach is highly significant for organizations that require sensitive information to be protected while offering specific individuals the capability to work with core data.

With field-level security, right personnel can access only specific information, and this will save sensitive information while allowing companies to comply with strict industry regulations such as GDPR (General Data Protection Regulation) for Europe, HIPAA (Health Insurance Portability and Accountability Act) for Health Services, and CCPA (California Consumer Privacy Act) for United States. Administrators can configure the field-level security for Objects from a Profile or Permission Set (see Figure 2-3).

All these requirements require protection of personal or confidential information, and access to certain data should be given only to a few who must have access to the same. Field-level security makes Salesforce a key player in enforcing all these standards and helps companies be better held accountable for the protection of customer information.

Field-level security preserves the privacy of data while complementing operations by making operational productivity better. It reduces clutter to the user by showing fields of information that are irrelevant to daily tasks by giving customized data views.

For example, a sales representative must be able to view contact details of customers but has no access to sensitive financial information such as credit scores or social security numbers. Similarly, a customer service representative who only needs to view a case record, may only require visibility on fields relevant to customer services, and not for details on internal metrics or account history. This makes the process of eliminating distractions from the focus of the users, thus enhancing productivity and avoiding errors that may be encouraged by accessing unnecessary information.

Of course, field-level security can be considered an important aspect that ought to be preserved as dictated by internal data governance requirements. Data access is handled with relatively fair accuracy in most companies, protecting not only the details of customers but also preventing internal data leakage and other malpractices involving sensitive corporate data. By controlling the visibility of data, companies protect information of this sort as strategic planning or financial metrics to only those who are authorized in the organizations.

Field-level security therefore enables the flexibility of dynamic organizations. This means that whenever the needs of the business change, administrators can have an adaptation at the field level based on the role change for compliance or other organizational goals.

For example, with the increasing scale of a business to a new region with specific regulations on privacy, the administrators may adjust the field-level security settings to fit this requirement. Adaptability is priceless as it allows Salesforce to scale and move with the organization without compromising data security.

In a nutshell, field-level security is a very strong tool in any Salesforce environment because it allows organizations to do the following:

- Protect sensitive information precisely by managing what fields to show and not to show

- Comply with data privacy regulations by putting restrictions on regulated data fields access

- Tailor data access according to the roles of each user and get an added edge in productivity

- Guard proprietary information and uphold the standards of internal data governance

- Let data visibility controls continue to flow without distortion as the organization evolves and grows

Field-Level Security enables organizations to confidently manage data access across various departments and roles, ensuring that each user interacts only with the information relevant to their responsibilities. Organizations can thus maintain a delicate balance between privacy, regulatory compliance, and user productivity by setting which fields are to be made invisible or not editable. This way, sensitive data stays protected even in a shared data environment. Organizations can hence address complex business and regulatory requirements by field-level security without giving away Salesforce's flexibility or usability and have a secure, adaptable platform in tune with dynamically shifting operations and compliance needs.

Field-Level Security Requirements in Real-Life Scenarios

Field-level security controls which specific fields on a record users can view or edit, enabling finer control beyond object access. In real-life scenarios, this is crucial for protecting sensitive information—such as financial data or personal details—while still allowing users to work effectively with the data they need.

Financial Services Securing Client Financial Data

A large financial services organization uses Salesforce to manage their relationships and interactions with its customers. While account managers require access to generic information such as the contact details of the customers, for example, and the history of an account, a few fields holding sensitive information such as credit scores, tax identification numbers, and annual income should be restricted by regulatory and privacy compliance requirements.

Fields containing such sensitive financial information would be locked down only for compliance officers and senior managers to view, while fields containing the same information the customer support agent or the junior account manager sees should be locked down as read- or write-only, respectively. It will allow administrators to lock down sensitive fields based on a job role, hence supporting compliance with privacy requirements but provides no more information to the different team members than is needed for them to get access.

Health Care Organization—HIPAA Compliance of Patient Records

A healthcare organization stores and manages patient data consisting of contact information, treatment history, and other health information in Salesforce Health Cloud. Whereas the doctor and nurse may need access to the full medical history of the patient, administrative staff would require basic details only for scheduling appointments and billing. Access to PHI is restricted unless absolutely needed when accessing PHI for treatment of patients even though that access is strictly regulated under HIPAA.

The access to health information, which is nonsensitive in nature, for example, diagnosis or treatment-related information, should only be accessible and amendable by licensed medical staff. Only administrative staff should view the fields containing nonsensitive information like contact information or a date of an appointment. This kind of field-level security ensures HIPAA compliance, protects the privacy of patients, and allows all these various types of roles within the organization to have customized access to the same data.

Best Practices for Using Profiles and Permission Sets

The proper assignment of permissions is essential to maintain a secure, scalable, and efficient Salesforce environment. Profiles and Permission Sets must be used strategically to design an access control structure that is structured, meets business requirements while avoiding the pitfalls of administration.

Follow Role-Based Design for Profiles

Start by examining your organization's structure and determining key roles in user organizations.

- **Map Organizational Roles to Profiles**: Map Profiles directly to clearly defined responsibilities—such as Sales, Support, or Operations—so that profiles throughout the organization have consistent permissions

- **Do Not Over-Specialize Profiles**: Over-specialized Profiles are inefficient. Instead of creating separate Profiles for all Sales Representatives working in different territories, build one Profile and apply Permission Sets to cover unique regional requirements

- **Scalability**: Create Profiles with a future-ready mindset. Anticipate organizational changes, such as new roles or shifts in responsibilities, and create Profiles that would adapt easily to these changes without significant rework

Maximize Permission Sets for Granular Access

Use Permission Sets for temporary or specialized access to enable projects or initiatives spanning multiple teams involved.

- **Allow Cross-Functional Collaboration**: If Marketing and Sales teams are working together on a campaign, a Permission Set would allow shared views into campaign information without having to change Profiles.

- **Simplify Temporary Access**: Allowing access for short-term needs is easily done with Permission Sets. For example, a consultant can be given a Permission Set for specific project-related data access that can be removed upon completion of that project.

- **Enable Advanced Features**: Use Permission Sets to manage access to such advanced features as API integration, developer tools, or third-party apps without automatically making these permissions the standard for a user group.

Organize Permission Sets on Purpose and Functionality

Group permissions logically according to tasks, like "Data Management" or "Customer Interaction Tools." This will make it easier to assign and track permissions across your users.

- **Do not Overlap Permissions**: In designing Permission Sets, avoid overlapping permissions by ensuring that each serves a different purpose. An example would be not duplicating the same permissions on several sets applied to one user.

- **Use Descriptive Names for Permission Sets**: Describe the purpose and scope of a Permission Set and assign it a name that reflects this. Instead of "Permission Set 1," the name "Advanced Case Management" would be more descriptive.

- **Leverage Permission Set Groups to Simplify Complex Assignments**: Group related Permission Sets together through Permission Set Groups for easier management. For instance, a "Field Service Agent" could group together permissions for Work Orders, Service Appointments, and Inventory Tracking.

- **Simplify Onboarding**: Automatically provide necessary tools to new users with complex roles through preconfigured Permission Set Groups, reducing the time to set up.

- **Monitor Group Usage**: Regularly review Permission Set Groups against current organizational needs. Update them as needs change or roles evolve.

Automate Permission Management

Set up automated processes for assigning or revoking permissions when there are roles changes—for example, promotions, transfers to another department.

- **Expiration Date**: Assign temporary permissions with an expiration date so they are automatically revoked after the specified period (Figure 2-6)

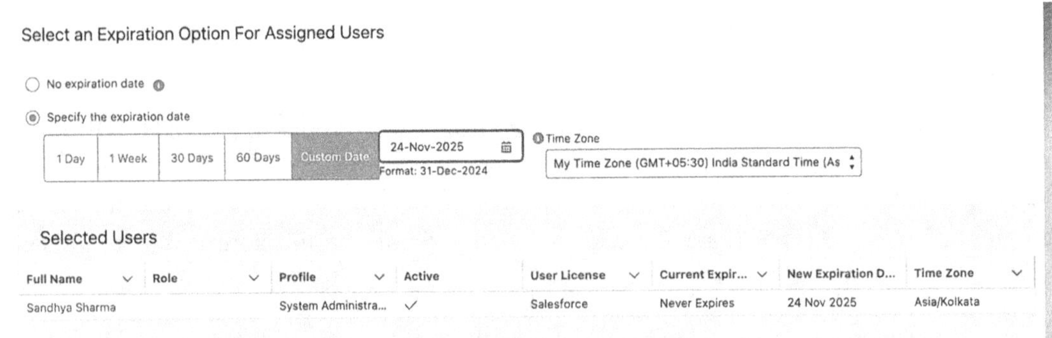

Full Name	Role	Profile	Active	User License	Current Expir...	New Expiration D...	Time Zone
Sandhya Sharma		System Administra...	✓	Salesforce	Never Expires	24 Nov 2025	Asia/Kolkata

Figure 2-6. *Permission set expiration setting user interface*

- **Review and Revise Permissions**: Routinely conduct scheduled audits to look at profiles and permission sets. Remove any unused or outdated configurations and clean them up to have the least amount of complexity in your systems

- **Engage with Stakeholders**: Work closely with the heads of departments to ensure that access permissions are still meeting business needs and workflows

- **Analyze Permission Assignments**: Identify users with conflicting or excessive permissions and then correct

Apply Documentation and Governance

Educating Salesforce administrators on best practice governance, and around permission management will be an ongoing process.

- **Maintain a Central Permission Log**: Record all Profiles, Permission Sets, and Permission Set Groups in a central location. Describe their business purpose and the associated roles and dependencies

- **Develop Governance Policies**: Clearly outline policies related to Profile and Permission Set creation, assignment, and retirement. Ensure all administrators use these guidelines

Challenges and Solutions

The solutions to the following challenges proactively ensure a permission management framework that is secure, flexible, and aligned with organizational goals. The power of Salesforce tools, integrated with strategic planning, allows administrators to balance these competing demands between security, scalability, and usability.

Challenge 1: Over-Permissioning

Problem: Users are given too much permission, creating a culture of too many permissions and too little integrity in data.

Solution: Define strict rules for each user's role. Use Permission Sets only where absolutely necessary for added permissions to a user. Regularly review all users' permissions to see if they have grown out of their current set.

Challenge 2: Infrastructural Maladjustment for Profiles

Problem: An inflexible, static definition of Profiles is not able to accommodate organizational changes that come from new roles or teams.

Solution: Scalable Profile Designs. Make higher-level Profiles for key roles and Permission Sets for access requirements. That will let the system grow along with your organization.

Challenge 3: Deliberate Access Over-Granting

Problem: Temporary permissions are often forgotten, leaving users with excess access long after it is needed.

Solution: Implement processes for tracking and revoking temporary access. Use tools like expiration dates for Permission Sets or create automated workflows for managing access life cycles.

Challenge 4: Lack of Alignment with Business Processes

Problem: Permissions do not map to the actual workflows of departments, leading to inefficiencies or user frustration.

Solution: Engage stakeholders in understanding their needs before designing Profiles and Permission Sets. Continuously revisit configurations to ensure they support evolving business processes.

Challenge 5: Inadequate Role Documentation

Problem: Roles and permission are not clearly documented, which hampers troubleshooting and auditing.

Solution: Invest in intensive documentation of all Profiles, Permission Sets, and Permission Set Groups. Include the reasoning behind how they were created and any applicable guidelines for usage.

Brainstorming

1. **What is the primary purpose of Profiles in Salesforce?**

   A. Grant specific permissions to individual users

   B. Define baseline permissions for user groups

   C. Temporarily assign access

   D. Manage sharing rules

   Answer: B

2. **Which feature is NOT controlled by Profiles?**

   A. Login IP ranges

   B. Object permissions

   C. Temporary permissions

   D. Field-level security

   Answer: C

3. **What is a key use case for Permission Sets?**

   A. Configuring default record types

   B. Granting access to individual users for specific tasks

   C. Enforcing IP restrictions

   D. Setting baseline permissions for user groups

   Answer: B

4. **Which feature is unique to Profiles but not Permission Sets?**

   A. Granting field-level access

   B. Controlling login IP ranges

   C. Extending object-level permissions

   D. Providing access to custom applications

   Answer: B

5. **How can Permission Set Groups simplify permission management?**

    A. By creating new Profiles automatically

    B. By grouping related Permission Sets for easier assignment

    C. By granting access to all users by default

    D. By enforcing record-level security

    Answer: B

6. **Which of the following can be managed using Permission Sets?**

    A. Login IP restrictions

    B. Field-level security

    C. Default record types

    D. Sharing rules

    Answer: B

7. **What is the recommended best practice for using Profiles?**

    A. Use them to define baseline permissions for user groups

    B. Assign unique Profiles to each user

    C. Use Profiles for temporary permissions

    D. Avoid using Profiles altogether

    Answer: A

8. **What is a common challenge when using Profiles?**

    A. They lack field-level security options.

    B. They cannot grant object-level access.

    C. Creating too many Profiles can lead to "Profile sprawl."

    D. They override role hierarchy settings.

    Answer: C

9. **Which of the following is an example of a temporary permission?**

    A. Granting access to a specific object for a short-term project

    B. Defining baseline access for a department

    C. Setting default record types for users

    D. Assigning login hour restrictions

    Answer: A

10. **What is the primary benefit of Permission Sets?**

    A. They simplify organization-wide settings.

    B. They allow for exception-based access without creating new Profiles.

    C. They enforce login IP ranges.

    D. They assign default tab visibility.

    Answer: B

11. **When should you use Permission Sets instead of Profiles?**

    A. To grant baseline permissions

    B. To extend permissions for specific users

    C. To set up login hour restrictions

    D. To enforce field-level security across all users

    Answer: B

12. **Which best practice applies to Permission Set Groups?**

    A. Avoid grouping related Permission Sets

    B. Use them to consolidate permissions and simplify assignments

    C. Assign Permission Set Groups to all users indiscriminately

    D. Only use them for field-level security

    Answer: B

13. What is a limitation of using Permission Sets?

A.   They cannot grant field-level security.

B.   They do not support baseline permissions.

C.   They do not enforce IP or login hour restrictions.

D.   They override Profiles by default.

Answer: C

14. What is the recommended way to avoid Profile sprawl?

A.   Create highly specific Profiles for each user role

B.   Use broad Profiles and Permission Sets for exceptions

C.   Use Permission Sets exclusively without Profiles

D.   Assign each user a unique Profile

Answer: B

15. How often should Profiles and Permission Sets be audited?

A.   Annually

B.   Only when new roles are created

C.   Regularly, to ensure permissions are up-to-date

D.   Never, once they are set

Answer: C

16. What is the purpose of documenting Profiles and Permission Sets?

A.   To ensure login hour restrictions are properly configured

B.   To simplify future audits and debugging

C.   To enforce role hierarchy sharing rules

D.   To automate user onboarding

Answer: B

17. **Which of the following is NOT a key feature of Permission Sets?**

    A.   Granting temporary permissions

    B.   Defining baseline access

    C.   Extending object-level access

    D.   Adding permissions for specific users

    Answer: B

18. **What is one of the key advantages of Permission Set Groups?**

    A.   They override Profile settings automatically.

    B.   They group related Permission Sets to streamline
         assignments.

    C.   They enforce field-level security across all users.

    D.   They replace the need for Profiles entirely.

    Answer: B

19. **Which feature is exclusive to Profiles?**

    A.   Setting default record types

    B.   Defining login IP ranges

    C.   Extending object-level access

    D.   Granting permissions to individual users

    Answer: B

20. **When should you use a Profile instead of a Permission Set?**

    A.   To grant additional permissions temporarily

    B.   To define baseline permissions for a department

    C.   To configure login hour restrictions

    D.   To assign API access to specific users

    Answer: B

21. **Which is the best use case for Permission Sets?**

    A.   Defining baseline permissions for all users

    B.   Assigning temporary access for a specific task

    C.   Configuring login IP ranges for user groups

    D.   Managing organization-wide settings

    Answer: B

22. **What is the primary drawback of creating too many Profiles?**

    A.   It limits the use of Permission Sets.

    B.   It can lead to administrative overhead and confusion.

    C.   It enforces stricter access controls.

    D.   It simplifies user management.

    Answer: B

23. **Which is a scenario where Permission Set Groups are beneficial?**

    A.   Granting login IP restrictions

    B.   Managing complex permissions across multiple roles

    C.   Enforcing sharing rules

    D.   Replacing the need for Profiles

    Answer: B

24. **Which feature can be configured only using Profiles?**

    A.   Field-level security

    B.   Login IP ranges

    C.   Object-level permissions

    D.   API access

    Answer: B

25. **What is one limitation of Permission Sets?**

    A.   They do not support field-level security.

    B.   They cannot replace Profiles for baseline permissions.

    C.   They enforce login hour restrictions.

    D.   They override existing role hierarchies.

    Answer: B

26. **Which is NOT a recommended best practice for managing Profiles and Permission Sets?**

    A.   Document permissions thoroughly

    B.   Regularly audit and update permissions

    C.   Assign each user a unique Profile

    D.   Use broad Profiles and Permission Sets for exceptions

    Answer: C

27. **Which challenge does Permission Set Groups address?**

    A.   Lack of field-level security

    B.   Profile sprawl due to excessive customization

    C.   Enforcing login IP ranges

    D.   Defining sharing rules

    Answer: B

28. **How does using Permission Sets improve flexibility?**

    A.   By replacing Profiles entirely

    B.   By granting object-level permissions to specific users

    C.   By simplifying login IP restrictions

    D.   By enforcing role hierarchy rules

    Answer: B

29. **What is one advantage of documenting Profiles and Permission Sets?**

   A.   It allows automation of user onboarding.

   B.   It prevents unintended access overlaps.

   C.   It simplifies enforcement of login restrictions.

   D.   It ensures proper configuration of sharing rules.

   Answer: B

30. **Which permissions cannot be granted using Permission Sets?**

   A.   Object-level access

   B.   Field-level security

   C.   Login IP ranges

   D.   Access to custom applications

   Answer: C

31. **What is one best practice for assigning Permissions?**

   A.   Always use Permission Sets for baseline access

   B.   Avoid granting overlapping permissions

   C.   Assign temporary permissions using Profiles

   D.   Use Permission Sets exclusively

   Answer: B

32. **What is the best way to manage access for a temporary project?**

   A.   Create a new Profile for the project

   B.   Assign a Permission Set for the project's duration

   C.   Use login IP restrictions

   D.   Modify role hierarchy settings

   Answer: B

33. **Which is a feature unique to Profiles?**

    A.   Extending field-level security

    B.   Setting login hour restrictions

    C.   Assigning custom tabs

    D.   Granting object-level permissions

    Answer: B

34. **How does auditing Permission Sets improve security?**

    A.   By reducing Profile sprawl

    B.   By consolidating overlapping permissions

    C.   By automating user access

    D.   By simplifying role hierarchies

    Answer: B

35. **What is the role of Permission Set Groups?**

    A.   To replace Profiles for baseline permissions

    B.   To group related Permission Sets for easier management

    C.   To enforce sharing rules

    D.   To assign login hour restrictions

    Answer: B

36. **When should you prefer using a Profile over a Permission Set?**

    A.   To grant temporary access for a task

    B.   To enforce login IP restrictions

    C.   To extend permissions for specific users

    D.   To manage record-level security

    Answer: B

37. **Which of the following challenges does regular permission audits address?**

    A. Lack of field-level security

    B. Overlapping permissions causing unintended access

    C. Enforcing login hour restrictions

    D. Assigning baseline permissions

    Answer: B

38. **What is a common mistake when managing Permissions?**

    A. Using Permission Sets for temporary access

    B. Creating too many specific Profiles

    C. Documenting permission assignments

    D. Consolidating Permission Sets

    Answer: B

39. **Which is NOT a function of Permission Sets?**

    A. Granting temporary project access

    B. Extending baseline permissions

    C. Assigning login IP restrictions

    D. Enabling access to specific apps

    Answer: C

40. **Which is a recommended way to manage Permissions for a small team?**

    A. Assign a unique Profile to each team member

    B. Use a broad Profile and apply specific Permission Sets

    C. Create a custom sharing rule for each user

    D. Enforce login hour restrictions for the team

    Answer: B

41. How does Permission Set Groups help in managing user access?

A.   They override all Profile restrictions.

B.   They consolidate related Permission Sets for simplified management.

C.   They automate the enforcement of sharing rules.

D.   They replace Profiles entirely.

Answer: B

42. What is a key benefit of using Permission Sets for exceptions?

A.   They replace Profiles for baseline permissions.

B.   They minimize the need for additional Profiles.

C.   They enforce login IP restrictions.

D.   They automate field-level security updates.

Answer: B

43. Which of the following is a limitation of Permission Sets?

A.   They cannot be assigned to multiple users.

B.   They cannot enforce login hour restrictions.

C.   They override Profile permissions.

D.   They replace role hierarchies.

Answer: B

44. When should Permission Set Groups be used?

A.   To replace all Profiles in the system

B.   To enforce login IP restrictions

C.   To manage complex permissions across related roles

D.   To assign baseline permissions to users

Answer: C

45. **Which feature is unique to Profiles but not available in Permission Sets?**

A. Assigning access to custom objects

B. Enforcing login hour restrictions

C. Granting API access

D. Configuring field-level security

Answer: B

46. **What is a common challenge when managing multiple Permission Sets?**

A. Lack of flexibility in granting object-level permissions

B. Overlapping permissions leading to unintended access

C. Inability to assign users to multiple roles

D. Difficulty enforcing sharing rules

Answer: B

47. **Which practice helps prevent "Profile sprawl"?**

A. Assigning unique Profiles to each user

B. Using broad Profiles with Permission Sets for exceptions

C. Creating a Profile for each department

D. Consolidating all access into one Profile

Answer: B

48. **What is the main benefit of documenting Profiles and Permission Sets?**

A. It simplifies role hierarchy management.

B. It ensures clarity during audits and debugging.

C. It automates login IP range enforcement.

D. It reduces the need for Permission Set Groups.

Answer: B

49. Which is NOT a best practice for managing permissions?

A.   Regularly auditing Permission Sets and Profiles

B.   Using Permission Set Groups for complex permissions

C.   Creating very specific Profiles for small groups

D.   Documenting who has access to what and why

Answer: C

50. How can overlapping permissions be minimized?

A.   By consolidating Permission Sets into Permission Set Groups

B.   By assigning each user a unique Profile

C.   By using login hour restrictions

D.   By removing object-level access entirely

Answer: A

51. What is the primary purpose of Permission Sets?

A.   To define baseline permissions for large groups

B.   To grant specific permissions without altering Profiles

C.   To replace the role hierarchy model

D.   To enforce IP restrictions for certain users

Answer: B

52. What is a key advantage of using Permission Set Groups?

A.   They replace Profiles entirely.

B.   They simplify management by grouping related Permission Sets.

C.   They allow setting login IP ranges.

D.   They enforce role hierarchy settings.

Answer: B

53. **When is it better to use a Permission Set rather than creating a new Profile?**

    A.   When granting login hour restrictions

    B.   When assigning access to a temporary project

    C.   When defining baseline permissions for a team

    D.   When enforcing object-level access for all users

    Answer: B

54. **Which feature is exclusively managed by Profiles?**

    A.   Assigning apps to users

    B.   Login IP restrictions

    C.   Granting field-level security

    D.   Enabling object-level permissions

    Answer: B

55. **What is the recommended approach for managing temporary access?**

    A.   Creating a new Profile for the temporary access

    B.   Using a Permission Set for the specific period

    C.   Enforcing sharing rules for the duration

    D.   Modifying role hierarchies temporarily

    Answer: B

56. **Which is NOT a benefit of Permission Sets?**

    A.   Extending access without altering Profiles

    B.   Simplifying management of temporary permissions

    C.   Assigning login IP ranges

    D.   Enabling fine-grained control for individual users

    Answer: C

57. How do Permission Sets provide flexibility?

A. By replacing Profiles for baseline permissions

B. By allowing specific permissions to be added without altering Profiles

C. By enforcing login hour restrictions for individual users

D. By overriding sharing rules

Answer: B

58. What is one challenge addressed by Permission Set Groups?

A. Enforcing sharing rules

B. Managing complex overlapping permissions

C. Assigning login IP restrictions

D. Automating field-level security settings

Answer: B

59. Which is a best practice for managing Permissions?

A. Using unique Profiles for every team

B. Documenting and regularly auditing permissions

C. Granting baseline permissions exclusively through Permission Sets

D. Replacing Profiles with Permission Set Groups

Answer: B

60. What is one limitation of Permission Sets?

A. They cannot replace Profiles for baseline access.

B. They do not support object-level permissions.

C. They override Profile permissions entirely.

D. They cannot be assigned to multiple users.

Answer: A

CHAPTER 3

Role Hierarchies and Organization-Wide Defaults

Role hierarchies are core to the visibility and access model of Salesforce. They control data access in alignment with the reporting structure of any organization. It means, it ensures that the person placed in a higher role possesses visibility to the records owned by users in the subordinate roles. This technique gives administrators a flexible, scalable approach to manage record access across different objects while protecting data with robust security. In this chapter, we'll explore the concept of role hierarchies, showing how they regulate access to records, how they function, their advantages and limitations, and some practical application strategies.

How Role Hierarchies Control Record Access

Role hierarchies are an important tool for managing access to records in Salesforce. They help in aligning access with organizational structures, thus simplifying administration and enhancing security. When combined with OWDs, sharing rules, and other mechanisms, role hierarchies provide a robust framework for controlling data visibility in a scalable and efficient manner. Administrators should design and manage role hierarchies thoughtfully, balancing simplicity with the flexibility to meet complex business requirements.

© Sandhya Sharma 2026

S. Sharma, *Unlocking Salesforce Data*, https://doi.org/10.1007/979-8-8688-2305-3_3

Basic Understanding of Role Hierarchies

Essentially a role hierarchy is the represenation of the organizational hierarchy wherein roles are assigned in conformity with reporting relationships at work. Through the platform of Salesforce, administrators can set out these roles and their respective relationships, creating a model where access to records is determined or inherited upward. When a role hierarchy is used and a user is assigned a role, that role helps determine access to records (Figure 3-1).

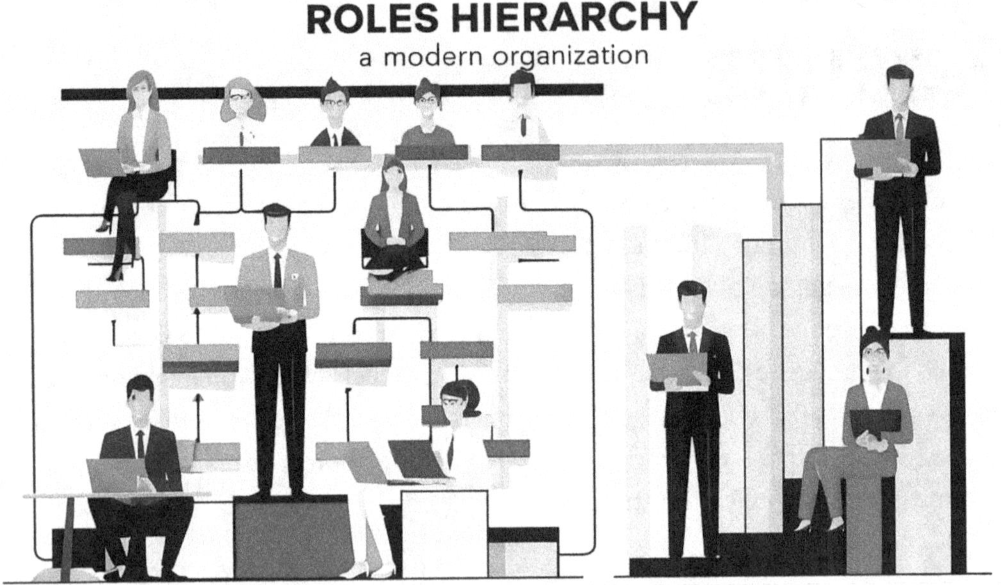

Figure 3-1. *A role hierarchy is a tree that signifies the chain of an organization*

Key Features include the following:

- Roles higher up in the hierarchy inherit access to the records owned by users lower down in the role hierarchy.

- Role hierarchies do not supplant object-level permissions set in profiles or permission sets; rather, they supplement those permissions.

- They are vertical in nature; access flows upward in the hierarchy.

Role Hierarchies at Work

Role hierarchies interact with the broader Salesforce security model to manage record visibility. Here's how they work:

- **Record Ownership**: Each Salesforce record has a designated owner, and record ownership provides the baseline access to a record.

  The owner is usually the user who created the record, but ownership can be reassigned.

- **Inheritance of Access**: Access is automatically inherited by a user higher in the role hierarchy to records owned by users lower in the hierarchy. For example, if a Sales Representative establishes a lead, their manager (e.g., Regional Sales Manager) may access that lead for viewing/editing based on the role hierarchy.

- **Interaction with Organization-Wide Defaults (OWDs)**: Role hierarchies rely on the OWDs for an object. When OWDs are set to "Private" or "Read Only," the role hierarchies provide access to the record to the users further up in the hierarchy. If OWDs are "Public Read/Write," role hierarchies do not introduce additional access since records are already accessible to everyone.

- **Granular Access Control**: Role hierarchies introduce an additional layer of control over object access. For instance, Read Only: Users higher in the hierarchy can view records but cannot make changes. Read/Write: Subordinates' records can be read and written by the users.

Use Cases for Role Hierarchy Examples

To understand how role hierarchies control access to records, consider the following real-world examples:

- **Sales Organization Scenario**: An organization has a Sales Representative role reporting to a Sales Manager, who in turn reports to a Regional Sales Director.

- **Role Hierarchy**: Sales Representative (lowest level); Sales Manager (middle level); Regional Sales Director (highest level)

- **Access Control**: A Sales Manager can view and edit all opportunities owned by their Sales Representatives.

  The Regional Sales Director will have access to all the opportunities of all the Sales Managers and Representatives within their region.

- **Customer Support**: The Support Representative is the lowest role, and they work individually on cases. However, one Support Manager manages many of these representatives.

- **Role Hierarchy**: Support Representative—lowest; Support Manager—highest

- **Access Control**: The Support Manager can access all cases under their representatives to follow through on the progress made and completion.

- **HR Department**: HR specialists maintain employee records, and all HR activities are monitored by the HR Manager.

- **Role Hierarchy**: HR Specialist (most junior); HR Manager (more senior)

- **Access Control**: The HR Manager can view all employee records, whereas the HR Specialists can only view the records they are responsible for.

Benefits of Role Hierarchies

Role hierarchies provide various benefits that are imperative components of the Salesforce data security model:

- **Alignment with Business Structures**: They reflect the organization reporting relationships, thus naturally easy to define and control.

- **Scalability**: Role hierarchies can adapt to changes in organizational structures, such as new roles or teams.

- **Lower Administrative Overhead**: Inheritance of access means that the higher-level users do not require additional sharing rules or manual configurations to view the records of their subordinates.

- **Security**: Role hierarchies complement OWDs and other sharing mechanisms to ensure a balance between access and security.

- **Collaboration**: Managers and team leads are allowed access to the records they require to manage their teams appropriately.

Limitations of Role Hierarchies

Although role hierarchies are powerful, they do come with limitations that the administrator has to keep in mind:

- **Vertical Access Only**: Role hierarchies provide access to records vertically. Horizontal sharing is not available across roles.

- **Effectiveness Is Tied to OWDs**: If OWDs are too lax, role hierarchies can turn redundant.

- **Static Structure**: Role hierarchies don't change and don't easily map to matrix or project-based structures.

- **Role Complexity with Large Organizations**: This becomes problematic with large organizations containing deep hierarchies: it is challenging to manage such a large number of roles and their interconnections.

Strategies for Role Hierarchies

To enable effective role hierarchies, administrators should adopt these strategies:

- **Define OWDs First**: First, set OWDs to "Private" or "Read Only" in those objects where data must be secured.

- **Use Meaningful Role Names**: Clearly define roles to represent their purpose and level in the hierarchy, like North Region Sales Manager.

- **Limit the Number of Levels**: The flatter the hierarchy, the less complex it is, and the better it performs.

- **Combine with Sharing Rules**: Sharing rules can extend access horizontally where needed to complement the vertical access that role hierarchies provide.

- **Regular Audits**: Review and update the hierarchy regularly to reflect the current organizational structure.

Configuring Organization-Wide Defaults

Organization-Wide Defaults (OWDs) are the bedrock of Salesforce's security and data-sharing model. They define the base access levels across records in an organization; more granular access control mechanisms, such as sharing rules, role hierarchies, and manual sharing are built on top of this level.

OWDs define the default organization-wide level of access all users have to records outside their ownership. They establish and enforce uniform and secure data visibility within organizations and place restrictions in such a way that any deviation requires override through other provisions.

The following are important characteristics of OWDs:

- **Baseline Security**: It determines the minimum level of access granted to users, which is automatically effective unless changed.

- **Object-Specific**: Every object has its own OWD setting, which can be customized to the needs of an organization.

- **Sharing Impact**: They determine how other sharing mechanisms, like role hierarchies and sharing rules, work.

- OWDs make sure that sensitive data is protected while allowing flexibility to meet diverse business requirements.

Why Are OWDs Important?

OWDs play a critical role in Salesforce's security model by balancing accessibility and protection. Without OWDs, it would be challenging to manage record visibility efficiently.

Benefits of OWDs include the following:

- **Security**: It ensures that sensitive data is protected from unauthorized access.

- **Scalability**: It provides a foundation that scales well in large organizations.

- **Customizability**: It allows administrators to configure access levels for each object based on specific business needs.

OWD Settings

Salesforce offers several standard and custom object OWD settings. Depending on the object type, the use case varies.

For standard and custom objects, the following are the types of OWD settings that can be applied:

- **Private**: Only the record owners and users above them in the role hierarchy can view or edit the records. Suitable for sensitive data such as financial records or HR information.

- **Public Read Only**: All users can see, but only the record owners and the people above them in the role hierarchy can make edits. Good for reference data.

- **Public Read/Write**: All users can read and write records. Suitable for collaborative environments, where security of the data is not a high concern.

- **Controlled by Parent**: Access is controlled by the parent object in a master-detail relationship.

This is common in use cases that involve tightly coupled data, such as invoices tied to accounts.

For specific objects, the following are the types of OWD settings that can be applied:

- **Campaigns**: These can be set to either "Private" or "Public Full Access," depending on the user roles.

- **Price Books**: Access can be limited to "No Access" or set to "View Only" for users who need to see it.

- **Users**: Always set to "Private" to protect sensitive employee information.

Configuring OWD in Salesforce

The setup of OWDs requires a planned procedure for the correct implementation against organizational policies and requirements.

1.   Open Setup and go to Security and click Sharing Settings (Figure 3-2).

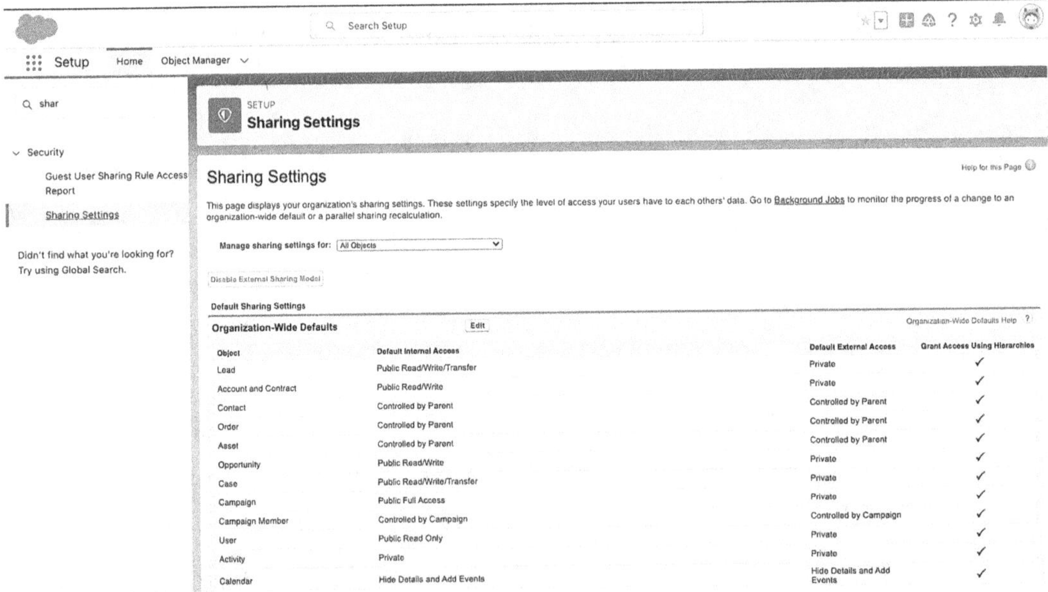

Figure 3-2. *Sharing Settings user interface*

2. Choose the desired object whose OWD needs to be set.

3. Choose the level of access as Private, Public Read Only, Public Read/Write, or Controlled by Parent.

4. Save and then verify the configuration by testing record access in various user roles.

Relationship of OWDs with Other Sharing Controls

OWDs are at the heart of the sharing model and can easily be deployed with other security controls, including the following:

- **Role Hierarchies**: This grants the ability to provide vertical access within an organization by sharing access with users that are ranked higher.

- **Sharing Rules**: Expand horizontal access with users or groups by using ownership rules or any defined criteria.

- **Manual Sharing**: Enables a record owner to share explicit access on an ad hoc basis.

- **Profiles and Permission Sets**: Control object-level access and field-level access but work in conjunction with OWDs to define record-level access.

Best Practices for Configuring OWDs

The following best practices outline key considerations and strategic approaches to effectively configure Organization-Wide Defaults (OWDs) for secure and scalable data access:

- **Start with Private**: Always begin with the most restrictive setting (Private) and gradually open access as required. This minimizes data exposure by default.

- **Evaluate Business Needs**: Understand the data sensitivity and collaboration requirements of your organization before finalizing OWDs.

- **Document Changes**: Maintain a history of configuration changes to better monitor decisions and aid in troubleshooting.

- **Test Thoroughly**: Periodically test record access to ensure configurations meet the desired outcomes.

- **Leverage Mechanisms**: Combine role hierarchies, sharing rules, and manual sharing to supplement OWD settings.

Use Cases for OWD Configurations

Below are sample use cases demonstrating how different OWD configurations address real-world data visibility and access control requirements:

- **Private OWD**: Applied to sensitive data, such as financial information or employee records. Role hierarchies and sharing rules can be used to further provide access.

- **Public Read Only OWD**: For reference data, such as a product catalog, where visibility is key but editing is limited.

- **Public Read/Write OWD**: Use for team-based projects, where everyone needs to be able to edit.

- **Controlled by Parent**: Ideal for child records where the master record controls it, such as invoices associated with accounts.

Challenges and Solutions in Managing OWDs

The following points highlight common challenges faced when managing OWD settings, along with practical solutions to address them effectively:

- **Overly Restrictive Settings**

  Problem: Users may find themselves unable to access necessary records.

  Solution: Use sharing rules or role hierarchies for increased access.

- **Overly Broad Configurations**

  Problem: It may inadvertently leak some sensitive information.

Solution: Use restrictive OWDs and open the access gradually.

- **Scaling Issues in Large Companies**

  Problem: Managing OWDs across multiple teams' objects will be difficult.

  Solution: The management of OWD can be simplified using fine-grained documentation and consistently followed naming conventions.

- **Meeting the Needs of Collaboration vs. Security Policies**

  Problem: Collaboration needs can oppose the security policies.

  Solution: Utilize the Public Read Only features along with controlled edit permissions for specific roles or teams.

Customizing Role Hierarchies for Complex Business Structures

Role hierarchies in Salesforce are part of its security and sharing model. This framework provides for the access of records through the reporting structure of an organization. When businesses grow and move toward more complex structures, role hierarchies also have to be adapted so that they work in compliance with complex workflows, geographic divisions, and compliance issues. This section delves into how to adapt role hierarchies effectively so that they satisfy the specific business needs, while not losing the security or efficiency.

Role Hierarchies Basics

Role hierarchies allow records to be accessed by allowing higher-level roles to see records owned by subordinate roles. They are closely linked to the OWDs and make record access flow vertically upwards in the hierarchy. This model simplifies record sharing inside a team or department but enforces access restrictions between unrelated teams.

Key principles include the following:

- **Vertical Access to Records**: Managers and users at higher levels can get direct access to the records of lower-level users.

- **Role-Based Visibility**: Access is granted based on the roles defined in the hierarchy, not the individual user.

- **Compliance with OWDs**: This role hierarchy complements OWDs, which will define the baseline level of access to records.

For instance, a sales team could have an organization structure such that a Regional Sales Manager can see all the opportunities owned by the Sales Representatives reporting to him in his region.

Complex Organizations Need Customization

Not all organizations have simple reporting structures. For instance, geographical distribution, matrix reporting, more than one business unit, and compliance necessitate role hierarchies customized. Role hierarchy customization addresses scenarios such as:

- **Geographic**: Access to data that relates specifically to their region or country.

- **Cross-Functional Teams**: Employees collaborating on projects require access to shared records.

- **Divisional Autonomy**: Different business units require isolation to their data but visibility by senior leadership to all parts of the globe.

- **Sensitive Data Protection**: Certain regulatory requirements demand restricted access to specific records.

Steps in Role Hierarchy Customization:

Role hierarchy customization is the alignment of the hierarchy with the structure and access requirements of the organization.

Step 1: Assess organizational needs

Start by understanding the company's reporting lines, operational workflows, and compliance requirements. Identify key stakeholders and their data access needs (Figure 3-3).

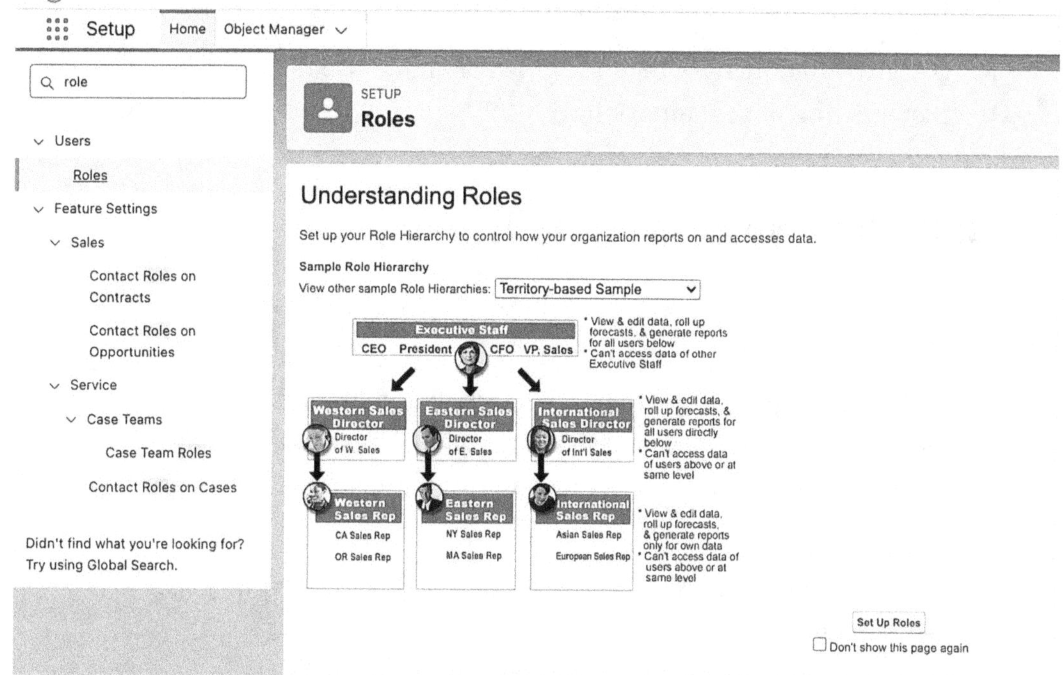

Figure 3-3. *Role hierarchy setup user interface*

Step 2: Design the hierarchy

Draft a role hierarchy that reflects the organization's structure. Define broad roles (e.g., executives, managers, individual contributors) and add subroles for granularity.

Step 3: Map access requirements

Identify which roles require access to certain data. Ensure the vertical access flow is logical and aligns with business requirements.

Step 4: In Salesforce

Utilize the Role Hierarchy feature within Salesforce to create and assign roles. Map users to roles based on their tasks.

Step 5: Test and validate

Thoroughly test to ensure that the hierarchy provides the right level of access. Confirm that records are only accessible to the required roles.

Step 6: Scalability optimization

Design the hierarchy with future growth in mind. Do not overly customize to prevent inflexibility.

Customization with Elaborate Techniques

In big complex organizations, role hierarchies usually call for creative approaches to uniquely tackle challenges.

- **Multi-Geographic Hierarchies**: For international companies, hierarchies of roles may be divided by region. Local managers may have access locally but can see globally from executives. Here is an example:

  CEO → Regional VP (Americas, EMEA, APAC) → Country Managers → Team Leads

- **Hierarchies of Roles in Cross-Functional Teams**: In matrix organizations, employees can have multiple managers depending on functions or projects. In such scenarios, sharing rules or manual sharing can complement role hierarchies.

- **Territory Management Integration**: Enterprise Territory Management complements role hierarchies within organizations with geographically defined operations. Overlapping or sharing territories does not impact the core hierarchy.

Tools Enhancing Role Hierarchy Flexibility and Efficiency

Though role hierarchies create a solid foundation for granting record access, there are also additional tools that can add to their flexibility and efficiency.

- **Sharing Rules**: Extend access horizontally by sharing records across roles or groups. For instance, let a marketing team be able to access campaign records owned by the sales team.

- **Manual Sharing**: Enable record owners to share access to individual users or roles for specific records. This is particularly useful for ad hoc collaboration.

- **Permission Sets**: Manage object- and field-level permissions separately from record-level access. Use permission sets to grant specific privileges without modifying the role hierarchy.

- **Field-Level Security**: Visibility or edit access can be limited to certain fields so that information is still protected even if records are shared.

Issues and Solutions in Role Hierarchies' Customization

Outlined below are common issues encountered while customizing role hierarchies, along with recommended solutions to maintain clarity and governance.

- **Complexity vs. Usability**: If the hierarchy becomes too complex, users may get confused and the administration is cumbersome. To prevent this, the structure should remain intuitive and include complementary sharing mechanisms for exceptions.

- **Handling Overlapping Responsibilities**: Cross-functional teams often blur the lines of traditional hierarchies. Use sharing rules or manual sharing to grant access without disrupting the hierarchy.

- **Maintaining Compliance**: Regulations may require restricted access to sensitive data. Combine private OWDs, role hierarchies, and field-level security to enforce compliance.

- **Scaling the Hierarchy**: Static hierarchies may not adapt well to organizational growth. Periodically review and update the structure to align with changing business needs.

Best Practices for Customizing Role Hierarchies

The following best practices provide guidance on designing and maintaining role hierarchies that align with organizational structure and data visibility needs:

- **Understand the Business Requirements**: Map the structure and workflows of the organization before designing the hierarchy.

- **Simplicity in Design**: The hierarchy should not have too much depth or complexity. The use of naming conventions improves clarity.

- **Add Sharing Rules and Permissions**: Rely on sharing rules, manual sharing, and permission sets to deal with exceptions and cross-functional access.

- **Test Thoroughly**: Validate the hierarchy with real-world scenarios to ensure it meets the business needs without sacrificing security.

- **Document and Communicate**: Keep clear documentation of the hierarchy and its rationale. Train users on their roles and access levels.

Use Cases for Customized Role Hierarchies

Below are practical scenarios illustrating how customized role hierarchies can support complex reporting lines and sharing requirements:

- **Multinational Corporations**: A global company uses regional hierarchies to limit access by geography, while senior executives have global oversight.

- **Retail Chains**: Store managers can view data specific to their stores, while regional managers oversee multiple stores, and corporate leadership has full visibility.

- **Healthcare Organizations**: Role hierarchies enforce HIPAA compliance by preventing access to patient records of those individuals who are not authorized, while allowing the doctors to view their cases assigned to them.

Best Practices for Efficient Role Hierarchy Design

Designing an efficient role hierarchy in Salesforce is critical for balancing record accessibility, security, and ease of administration. An efficient hierarchy reflects the organization's structure, supports the attainment of operational goals, and also provides scalability as the organization continues to grow. The following sections outline some best practices for administrators to help create and maintain such a hierarchy. Following these best practices will help the administrators design a hierarchy that enables business processes, enhances collaboration and keeps data secure.

Understand Organizational Structure and Needs

Before designing the role hierarchy, outline the organizational structure, including reporting lines, team divisions, and collaboration requirements. Identify key stakeholders, their roles, and the type of data access they need.

- **Assess Reporting Relationships**: The hierarchy should reflect the company's actual reporting relationships to make record access logical and intuitive.

- **Define Access Requirements**: Determine who needs to see records and how much access they require.

Begin with a Simple Structure

The role hierarchy should be as simple as possible yet meet the needs of the organization. Avoid overcomplicating the structure with unnecessary roles or depth.

- **Use Broad Roles First**: Define the top-level roles, "Manager," "Team Lead," and "Staff," and add more detail there as needed

- **Add Granularity When Needed**: Introduce subroles only when a particular business needs to enforce finer access control

Minimize Overlap

Overlapping responsibilities between roles can lead to confusion and inefficiencies. Clearly define the purpose of each role to ensure no two roles have redundant access.

- **Use Complementary Sharing Mechanisms**: Address overlapping access needs with tools like sharing rules, manual sharing, or permission sets instead of duplicating roles

Align with Organization-Wide Defaults (OWDs)

Role hierarchies, in turn, are integrated with OWDs to set record visibility. Be sure that the hierarchy complements the OWD settings to optimize performance.

- **Establish Appropriate OWDs**: When OWDs are set to "Private," the hierarchy will be crucial to sharing. For "Public" OWDs, the hierarchy may be only required in exceptional cases.

Scalability

The hierarchy must be adaptable to changes within an organization such as growth, mergers, or restructuring.

- **Anticipate Future Roles**: Leave space in the hierarchy for potential future roles or divisions

- **Minimize Hard-Coded Dependencies**: Use adaptive naming conventions and structures to help avoid major refactoring

Leverage Naming Conventions

Use obvious and descriptive naming conventions for the roles to promote clarity and simplify administration. For example, use names based on functions or reporting level, such as "Sales Manager—West" or "Support Specialist—Tier 2," use functional descriptions.

- **Include Business Unit Context**: For large organizations, include the business unit or region in the role name to avoid ambiguity

Regularly Audit and Refine

Periodically review the role hierarchy to ensure it remains relevant and efficient.

- **Remove Unused Roles**: Identify and deactivate roles no longer needed

- **Adjust for Changing Needs**: Update roles and reporting lines to reflect organizational changes

Train and Communicate

Educate users and administrators on the role hierarchy and its use. Ensure that they understand how their roles impact record access.

- **Provide Documentation**: Keep clear documentation of the hierarchy and associated access rules

- **Train Key Users**: Train managers and power users on optimal use of the hierarchy

Common Pitfalls to Avoid

Poorly designed access models can introduce risks and inefficiencies that impact both security and maintainability. Common pitfalls include the following:

- **Overcomplication**: Too many roles can make the hierarchy too hard to manage and understand.

- **Ignoring Exceptions**: When unique access requirements are not included, there are inefficient workarounds.

- **Security Neglected**: Overly broad roles can expose very sensitive data.

Brainstorming

1. **What is the primary function of a Role Hierarchy in Salesforce?**

   A. To define user permissions

   B. To control record-level access

   C. To assign profiles to users

   D. To configure field-level security

   Answer: B

2. **Organization-Wide Defaults (OWDs) determine**

   A. Field-level access

   B. User login access

   C. Baseline record visibility

   D. Role assignment rules

   Answer: C

3. **Which OWD setting provides the highest level of restriction?**

   A. Public Read/Write

   B. Public Read Only

   C. Controlled by Parent

   D. Private

   Answer: D

4. **Role Hierarchies provide**

   A. Horizontal access to records

   B. Vertical access to records

   C. Field-level security

   D. User login settings

   Answer: B

5. **In a private OWD setting, record access is granted based on**

   A. Field-level security

   B. Role Hierarchies and sharing rules

   C. User licenses

   D. Profiles

   Answer: B

6. **How are Role Hierarchies related to OWDs?**

   A. They override OWDs entirely.

   B. They supplement OWDs to extend record access.

   C. They function independently of OWDs.

   D. They restrict access granted by OWDs.

   Answer: B

7. **Which role in a hierarchy has access to the records of all users below them?**

   A. The lowest-level role

   B. The top-level role

   C. Roles at the same level

   D. Only system administrators

   Answer: B

8. **What happens if OWD is set to Public Read/Write?**

   A. Role Hierarchies are ignored.

   B. All users can edit all records.

   C. Only users with roles can edit records.

   D. Field-level security takes precedence.

   Answer: B

9. **Role Hierarchies grant record access to**

   A.   Users at the same level only

   B.   Users directly above in the hierarchy

   C.   Users above and at the same level in the hierarchy

   D.   System administrators only

   Answer: B

10. **What is the default OWD setting for newly created objects?**

   A.   Public Read/Write

   B.   Public Read Only

   C.   Private

   D.   Controlled by Parent

   Answer: C

11. **Customizing Role Hierarchies is ideal for**

   A.   Flat organizational structures

   B.   Complex, multi-level business structures

   C.   Public record visibility

   D.   Small businesses with few users

   Answer: B

12. **Which sharing tool can complement Role Hierarchies to grant additional access?**

   A.   Permission sets

   B.   Field-level security

   C.   Sharing rules

   D.   Profiles

   Answer: C

13. **If a user's role changes, their record access is updated based on**

    A. Their old role

    B. Their new role

    C. Their profile settings

    D. Manual sharing

    Answer: B

14. **In a Role Hierarchy, access to records is**

    A. Restricted by default

    B. Explicitly granted to all users

    C. Inherited from users below in the hierarchy

    D. Dependent on profiles

    Answer: C

15. **Organization-Wide Defaults apply to**

    A. Specific users only

    B. An entire object

    C. Role-specific records

    D. Custom report types

    Answer: B

16. **The "Controlled by Parent" OWD setting means**

    A. Record access is inherited from the parent object.

    B. Users can always edit parent records.

    C. Field-level security overrides this setting.

    D. Only system administrators can configure access.

    Answer: A

17. **How do Role Hierarchies handle cross-functional team access?**

    A. They allow unrestricted access to all users.

    B. They rely on manual sharing for cross-functional access.

C. They automatically share records across teams.

D. They are complemented by sharing rules for cross-functional access.

Answer: D

18. **Which feature allows granular access for specific users without changing Role Hierarchies?**

A. Profiles

B. Permission sets

C. OWDs

D. Login IP ranges

Answer: B

19. **Which statement about Role Hierarchies is false?**

A. They can be customized to match organizational structures.

B. They allow for flexible horizontal sharing.

C. They enable vertical record access.

D. They work in conjunction with OWDs.

Answer: B

20. **What is the first step when configuring a Role Hierarchy?**

A. Define sharing rules

B. Map out the organizational structure

C. Assign profiles to users

D. Enable OWD settings

Answer: B

21. **When is it recommended to set OWD to "Public Read Only"?**

A. When all users should have edit access.

B. When records are for view-only purposes across the organization.

C. When Role Hierarchies need to be bypassed.

D. When using field-level security.

Answer: B

22. **Which of the following statements about Salesforce role hierarchies is incorrect?**

A. Users higher in the role hierarchy can view and edit records owned by users lower in the hierarchy.

B. Role Hierarchies automatically grant access to records based on ownership and sharing rules.

C. Role Hierarchies can be modified by users with the "Manage Users" permission.

D. Users in higher roles can access records owned by users in lower roles only if the "Grant Access Using Hierarchies" setting is enabled.

Answer: B

23. **What is one disadvantage of overly complex Role Hierarchies?**

A. They simplify record access.

B. They are difficult to maintain and manage.

C. They increase record visibility.

D. They eliminate the need for sharing rules.

Answer: B

24. **Which of the following best describes the interplay between Role Hierarchies and OWDs?**

A. Role Hierarchies replace OWDs.

B. Role Hierarchies expand access based on OWD settings.

C. OWDs override Role Hierarchies.

D. Role Hierarchies restrict OWD settings.

Answer: B

25. **A manager in the Role Hierarchy gains access to records owned by**

   A.   Users above them in the hierarchy

   B.   Users below them in the hierarchy

   C.   All users in the organization

   D.   Users outside the hierarchy

   Answer: B

26. **Organization-Wide Defaults provide access at which level?**

   A.   Object level

   B.   Field level

   C.   User level

   D.   Record type level

   Answer: A

27. **What happens when a record owner's role is higher in the hierarchy?**

   A.   The record becomes public.

   B.   The record is only accessible by the owner.

   C.   Users below the owner in the hierarchy lose access.

   D.   The owner's role automatically grants access to higher roles.

   Answer: D

28. **How can access be extended when OWD is Private?**

   A.   Profiles

   B.   Permission sets

   C.   Sharing rules or Role Hierarchies

   D.   Public groups

   Answer: C

29. **To restrict record visibility in Salesforce, which OWD setting is recommended?**

    A.   Public Read/Write

    B.   Public Read Only

    C.   Controlled by Parent

    D.   Private

    Answer: D

30. **What is the impact of enabling Role Hierarchies?**

    A.   Users at the bottom of the hierarchy gain access to all records.

    B.   Users higher in the hierarchy gain access to records owned by users below.

    C.   Role Hierarchies disable sharing rules.

    D.   Users lose access to their own records.

    Answer: B

31. **When OWD is set to Public Read Only, what additional configuration is needed to allow edits?**

    A.   Profiles

    B.   Role Hierarchies

    C.   Sharing rules or manual sharing

    D.   Permission sets

    Answer: C

32. **Which feature provides the foundation for Salesforce's record-level access?**

    A.   Profiles

    B.   Organization-Wide Defaults

    C.   Permission sets

    D.   Validation rules

    Answer: B

33. **If a Role Hierarchy is not aligned with the organization's structure, what might occur?**

    A. Reduced system performance

    B. Inconsistent record access

    C. Automatic updates to the hierarchy

    D. Locked OWD settings

    Answer: B

34. **The term "inheritance" in Role Hierarchies refers to**

    A. Transferring roles to other users

    B. Access granted to roles higher in the hierarchy

    C. Sharing records with external users

    D. Cloning roles for new users

    Answer: B

35. **What type of OWD setting is "Controlled by Parent" most commonly used for?**

    A. Accounts

    B. Contacts

    C. Custom objects with master-detail relationships

    D. Leads

    Answer: C

36. **Which is the most restrictive OWD setting?**

    A. Public Read/Write

    B. Public Read Only

    C. Controlled by Parent

    D. Private

    Answer: D

37. **Role Hierarchies are designed to**

   A. Replace profiles

   B. Provide vertical record access

   C. Restrict object-level permissions

   D. Create user groups for sharing

   Answer: B

38. **What is a key limitation of Role Hierarchies?**

   A. They only work for standard objects.

   B. They cannot grant access to users outside the hierarchy.

   C. They override all OWD settings.

   D. They require manual configuration for every user.

   Answer: B

39. **What should you do before designing a Role Hierarchy?**

   A. Assign users to profiles

   B. Map out your organization's structure

   C. Enable field-level security

   D. Set OWD to Public Read/Write

   Answer: B

40. **Sharing rules work alongside Role Hierarchies to**

   A. Restrict access to certain records

   B. Extend record access beyond the hierarchy

   C. Grant object-level access

   D. Assign custom field-level security

   Answer: B

41. **In a private OWD setting, Role Hierarchies**

   A.   Provide full access to all records

   B.   Restrict all access to records

   C.   Extend access to users higher in the hierarchy

   D.   Are disabled by default

   Answer: C

42. **What happens when OWD is set to Public Read/Write?**

   A.   Role Hierarchies are ignored.

   B.   All users can edit all records.

   C.   Access is determined by profiles only.

   D.   All users can view but not edit records.

   Answer: B

43. **The top-level role in a Role Hierarchy automatically has access to**

   A.   Only their own records

   B.   Records owned by users at lower levels

   C.   All records in the organization

   D.   Records shared manually

   Answer: B

44. **If a record owner leaves the company, how is record access affected in a Role Hierarchy?**

   A.   The records become inaccessible.

   B.   The records are reassigned to the next user in the hierarchy.

   C.   The records remain accessible to roles higher in the hierarchy.

   D.   The records are deleted.

   Answer: C

45. **OWD settings are most useful when**

   A. Role Hierarchies are not in place

   B. Defining baseline record visibility across the organization

   C. Configuring field-level security

   D. Managing external user access

   Answer: B

46. **In a Role Hierarchy, who inherits record access?**

   A. Users at the same level in the hierarchy

   B. Users below the record owner

   C. Users higher in the hierarchy

   D. Only system administrators

   Answer: C

47. **Which OWD setting allows the least amount of visibility?**

   A. Public Read/Write

   B. Public Read Only

   C. Controlled by Parent

   D. Private

   Answer: D

48. **What is the relationship between OWDs and sharing rules?**

   A. Sharing rules restrict OWDs.

   B. Sharing rules expand access defined by OWDs.

   C. Sharing rules override OWDs.

   D. Sharing rules operate independently of OWDs.

   Answer: B

49. **Which configuration is required for a hierarchical access model?**

A.   Permission sets

B.   Field-level security

C.   Role Hierarchies

D.   Custom report types

Answer: C

50. **In which scenario would you use "Controlled by Parent"?**

A.   When managing unrelated records

B.   When dealing with master-detail relationships

C.   For objects with many-to-many relationships

D.   For assigning custom permissions

Answer: B

51. **What should you do if Role Hierarchies are too rigid for your sharing needs?**

A.   Use profiles instead

B.   Rely on sharing rules or manual sharing

C.   Enable Public Read/Write OWD

D.   Configure field-level security

Answer: B

52. **If a new user is added to a role, their access to existing records is determined by**

A.   Their profile

B.   Their assigned role

C.   Manual sharing settings

D.   OWD configuration

Answer: B

53. What happens when OWD is set to Public Read Only?

A. All users can view and edit records.

B. All users can view records but cannot edit.

C. Access is restricted to record owners only.

D. Access is determined by profiles only.

Answer: B

54. Role Hierarchies automatically

A. Assign custom profiles to users

B. Allow access to records owned by users below a role

C. Restrict access to records for higher-level roles

D. Override sharing rules

Answer: B

55. What is the most efficient way to manage Role Hierarchies in large organizations?

A. Avoid creating them entirely

B. Keep them simple and aligned with business structures

C. Create custom roles for every user

D. Use field-level security instead

Answer: B

56. OWD settings apply to

A. All standard objects only

B. Both standard and custom objects

C. Field-level permissions

D. External users only

Answer: B

57. **How do Role Hierarchies affect reporting in Salesforce?**

    A. They do not affect reporting.

    B. They restrict access to dashboards.

    C. They enable visibility into records owned by subordinates.

    D. They override field-level security in reports.

    Answer: C

58. **Which OWD setting allows record access to depend on parent object settings?**

    A. Private

    B. Public Read/Write

    C. Controlled by Parent

    D. Public Read Only

    Answer: C

59. **Sharing rules are created to**

    A. Provide additional record access beyond Role Hierarchies

    B. Restrict record access

    C. Modify OWD settings

    D. Assign profiles to users

    Answer: A

60. **In a Role Hierarchy, users with the same role**

    A. Can always access each other's records

    B. Cannot access each other's records unless shared

    C. Automatically gain edit access to each other's records

    D. Have access based on OWD settings only

    Answer: B

CHAPTER 4

Sharing Rules and Manual Sharing

In the earlier chapters, we explored how **Role Hierarchies** and **Organization-Wide Defaults (OWDs)** form the bedrock of Salesforce's visibility model. Those settings ensure that access decisions start from a clearly defined baseline.

But in reality, business requirements are rarely that straightforward. You might have situations where:

- **Two Regional Teams** need to collaborate occasionally but maintain separate visibility most of the time.

- **Departments** require access to certain records based on specific field values—not just ownership.

- **Short-Term Projects** involve external partners who need a subset of records without broader access.

In these cases, relying solely on OWDs and role hierarchies either **overexposes** data or **blocks** legitimate collaboration.

This is where Sharing Rules and Manual Sharing step in. They introduce flexibility without dismantling the core security framework.

In this chapter, we will

- Break down **Criteria-Based** and **Owner-Based** Sharing Rules—when to use each, and how to design them

- Understand how to **Implement Effective Sharing Rules** for Teams and Departments

- Explore **Manual Sharing** for one-off access needs

- Discuss **Best practices** for building maintainable sharing models

© Sandhya Sharma 2026

S. Sharma, *Unlocking Salesforce Data*, https://doi.org/10.1007/979-8-8688-2305-3_4

- Highlight **Common Pitfalls** and performance considerations

- End with **Brainstorming Exercises** to connect these concepts to your own org's requirements

Criteria-Based and Owner-Based Sharing Rules

Imagine your role hierarchy as a **tree**. Information generally flows upward—managers see their subordinates' records, but not the other way around. This works for most internal reporting needs.

However, real-world collaboration often **cuts across the branches of that tree**. Two roles might be peers in the hierarchy with no natural visibility into each other's data—but still need to work together.

Sharing Rules bridge this gap by allowing admins to **extend record access across roles, public groups, or territories** without altering the hierarchy. This avoids unintended side effects of restructuring roles just to solve a single collaboration issue.

Owner-Based Sharing Rules

Grants access to records owned by users in specific roles, public groups, or territories.

When to Use:

- The deciding factor for access is **who owns the record**, not other attributes.

- Ownership patterns are **stable and predictable** over time.

Example—Regional Sales Collaboration:

A US sales team owns all opportunities for US-based customers. The European sales director wants visibility into these opportunities to identify cross-selling potential for multinational accounts.

Restructuring the role hierarchy to link both regions would cause unnecessary complexity—and possibly overexpose other records. Instead, an **owner-based sharing rule** allows the European sales director to view (or edit, if needed) opportunities owned by the US sales team.

How It Works in Practice:

1. Identify **the group** that owns the records

2. Identify **the group** that needs access

3. Set the **access level** (Read Only or Read/Write)

4. Activate the rule—Salesforce automatically applies it to matching records

Lessons Learned:

- **Public Groups** make these rules easier to maintain. If a person changes roles, you only update the group membership—not the rule itself.

- Review ownership patterns quarterly. Changes in territory alignment can unintentionally expand or restrict access if rules aren't updated.

Criteria-Based Sharing Rules

Grants access to records that meet specific **field-value conditions**, regardless of ownership.

When to Use:

- Access decisions depend on **record attributes** rather than who owns it.

- You need **dynamic** rules that update automatically when data changes.

Example—Product-Specific Case Visibility:

A tech support organization manages cases for multiple product lines—*Cloud Storage, Email Security,* and *Data Backup.*

Each product team should only see cases related to their product. Using a **criteria-based sharing rule**, the system grants access to all cases where Product__c = "Cloud Storage" to the Cloud Storage team.

If a case's product field changes, access updates automatically—without manual intervention.

Best Practices:

- Keep criteria **simple** to prevent performance slowdowns during recalculations.

- Review field dependencies. If your rule uses a field that changes frequently, it might trigger more sharing recalculations than necessary.

- Use **checkbox flags** or **picklist values** as criteria for cleaner logic.

Choosing Between Owner-Based and Criteria-Based

Selecting the appropriate sharing rule type depends on your organization's data ownership model, access complexity, and long-term scalability requirements. A summary of the differences between owner-based and criteria-based sharing rules is provided in Table 4-1.

Table 4-1. *Comparison of Owner-Based and Criteria-Based Sharing Rules*

Factor	Owner-Based	Criteria-Based
Access Trigger	Record owner	Field value
Best Use	Stable ownership patterns	Dynamic access needs
Example	Regional sales access	Department-specific HR records
Flexibility	Limited	High
Maintenance Effort	Low (if groups used)	Medium (criteria may need adjustments)

Key Takeaway: If access depends on **who owns the record** and that doesn't change often, use Owner-Based. If it depends on **what's in the record** and that can change, use Criteria-Based.

Implementing Effective Sharing Rules for Teams and Departments

Creating a sharing rule is easy. Creating a **good** sharing rule that **ages well** is harder. Poorly designed rules often

- Overexpose sensitive records

- Duplicate access paths that already exist elsewhere

- Cause performance degradation when data volumes grow

A disciplined approach can prevent these issues.

Step-by-Step Approach to Effective Rule Design

1. **Map the Collaboration Requirement**

    - Identify exactly **which teams** need to work together and **why**.

    - Example: The **Customer Success** team wants to see opportunities in the "Renewal" stage to prepare for client outreach.

2. **Define the Access Scope**

    - Share **only what's necessary** to achieve the business outcome.

    - If Customer Success only needs renewal opportunities, don't give them all opportunities.

3. **Select the Right Rule Type**

    - **Owner-Based** for predictable ownership groups

    - **Criteria-Based** for field-driven, changing access

4. **Choose the Access Level**

    - **Read Only** for review/reporting

    - **Read/Write** for joint execution

5. **Test in a Sandbox**

    - Simulate real user scenarios to confirm the rule works as intended.

6. **Document Everything**

    - Record the **business reason**, affected records, and expected outcomes.

Example—Customer Success and Sales Alignment

Scenario:

Customer Success wants to see opportunities that are in the *Renewal* stage. They don't own these opportunities; Sales does.

Solution:

A **criteria-based sharing rule** where Stage = Renewal grants access to the Customer Success role.

Why It Works:

- No changes to hierarchy

- No need to give access to unrelated opportunities

- Automatically updates when an opportunity moves into or out of the Renewal stage

Lessons Learned:

- Avoid chaining multiple sharing rules to solve one problem—consolidate where possible.

- Periodically review whether the "Renewal" stage definition has changed, which might affect the rule's logic.

Manual Sharing—Ad Hoc Access Control

Manual Sharing in Salesforce is the most **situational** and **granular** approach to granting access.

Unlike role hierarchies, OWDs, or sharing rules—which operate based on **system-defined logic**—manual sharing is **user-initiated** and record-specific.

Think of it as a **"door key"** you hand to someone temporarily.

It works perfectly for exceptional scenarios where creating a permanent sharing rule is overkill.

Key Characteristics:

- **Record-Specific**: Applies to one record at a time.

- **User-Driven**: Can be performed by the record owner or users with "Full Access" to that record.

- **Immediate Effect**: Changes take place instantly after the share is saved.

- **Reversible**: Access can be removed as easily as it is granted.

When to Use Manual Sharing

Manual sharing shines when

- **Temporary Access Is Needed**: for example, project-based collaboration

- **One-off Exceptions Arise**: without justifying a new sharing rule

- **Coverage During Absence**: replacing a record owner during leave

- **Sensitive Cases**: limiting exposure by sharing only the necessary record

Example—Merger and Acquisition Deal Room Access

During a merger, a senior executive needs access to a **specific Account record** that contains due diligence notes.

Instead of giving them visibility into **all accounts** via a sharing rule, the record owner manually grants access **only to that record**.

How Manual Sharing Works

The process typically involves

- Navigate to the record to be shared

- Use the **"Sharing"** button (visible only if OWD for that object is not Public Read/Write)

- Add users, roles, or groups

- Assign access level—**Read Only** or **Read/Write**

- Save

Note   If the OWD for an object is set to Public Read/Write, the Sharing button won't appear—because everyone already has full access.

Manual Sharing in Lightning vs. Classic

While manual sharing functionality exists in both Lightning and Classic, the **user interface** differs:

- In **Lightning Experience**, the Sharing button often appears under the **action menu** on the record page.

- In **Classic**, it appears as a related list called "Sharing".

Limitations of Manual Sharing

Despite its flexibility, manual sharing has notable constraints:

- **Ownership Dependency**: If the record owner changes, all manual shares are lost.

- **Not Universally Available**: Only works if the object's OWD is not set to Public Read/Write.

- **No Automation**: Cannot be triggered by workflows or flows directly (though Apex can manage manual share records).

- **Scalability Issues**: Impractical for mass sharing—use sharing rules for bulk access.

Best Practices for Manual Sharing

- Use manual sharing **sparingly**—only for exceptions.

- Document why a record was manually shared (can be done in Chatter or record notes).

- Train end users to understand when manual sharing is appropriate.

- For repeated exceptions, consider automating via Apex sharing or adjusting sharing rules.

Limitations and Considerations in Sharing Models

The Hidden Risks of Over-Sharing

Too much openness can backfire:

- **Compliance Breaches**: exposing regulated data (GDPR, HIPAA, etc.)

- **Competitive Risks**: one department seeing sensitive sales forecasts of another

- **Loss of Trust**—: internal disputes over inappropriate data visibility

Performance Implications

- Every sharing rule adds to the **sharing recalculation process**.

- Large orgs with **millions of records** can experience slowdowns if too many rules exist.

- Overlapping rules increase complexity and recalculation time.

Lesson Learned: At one large enterprise, a "share everything for collaboration" approach led to a **two-hour sharing recalculation** every time territories were updated—slowing down the business.

Maintainability Challenges

- Mergers, org restructures, and role changes require **sharing model updates**.

- Without proper documentation, old rules may linger long after they're needed.

- Layered complexity can lead to **"invisible" access paths** that are hard to audit.

Strategic Recommendations

- **Start Simple**—: the simplest model that meets the requirement is often the best.

- **Document Every Rule**—: include business justification, scope, and owner.

- **Review Quarterly**—: verify that all rules still serve a purpose.

- **Limit Criteria Complexity**: complex formulas slow performance.

- **Test Before Deploying**: especially in large-data-volume environments.

Exercises

Mapping Sharing Rules to Real Business Needs

Purpose: To ensure that every sharing decision aligns with business objectives—not just "because someone asked for it."

Discussion Starters

- Which business units in your org require **cross-team visibility**?

- Are there **exceptions** that repeat often enough to justify a rule?

- How does your **regulatory environment** shape sharing decisions?

- Which scenarios in your org **still require manual sharing**?

Thought Exercise

You are the Salesforce architect for **GlobalTech**, a multinational with

- **Three Regional Sales Teams** (Americas, EMEA, APAC)

- **One Central Marketing Team**

- **One Support Department** with four product lines

114

Challenge:

Design a sharing model that

- Allows marketing to see **closed-won opportunities** globally.

- Let's support teams only see cases for **their assigned product line**.

- Permits occasional ad hoc account sharing between regional sales managers.

Questions to Answer:

- Which requirements are permanent (sharing rules)?

- Which are temporary (manual sharing)?

- What risks do you foresee if rules are misconfigured?

Activity

1. **Create a Stakeholder Matrix**

   - List all key teams/roles.

   - Note their data needs and access level.

   - Flag high-risk data points.

2. **Draft a Rule-Access Map**

   - Match each access need to a sharing method.

   - Identify if the method is **owner-based**, **criteria-based**, or **manual**.

3. **Simulate a Change Event**

   - Imagine the company merges with another.

   - Review if your sharing model can handle new teams without massive redesign.

Brainstorming

1. **What is the main purpose of sharing rules in Salesforce?**

   A. To grant field-level security

   B. To extend record access beyond role hierarchy settings

   C. To configure page layouts

   D. To define user profiles

   Answer: B

2. **Which type of sharing rule grants access based on the record owner?**

   A. Criteria-based sharing rule

   B. Owner-based sharing rule

   C. Manual sharing

   D. Public group sharing

   Answer: B

3. **In Salesforce, criteria-based sharing rules grant access based on**

   A. The user's role hierarchy position

   B. Specific field values in a record

   C. The type of license assigned

   D. The profile permissions

   Answer: B

4. **Which sharing setting must be considered before creating sharing rules?**

   A. Profile settings

   B. Organization-Wide Defaults (OWDs)

   C. Login IP ranges

   D. Password policies

   Answer: B

5. **If a team needs access to records they don't own, what's the best approach without changing the role hierarchy?**

    A. Modify profiles

    B. Create a sharing rule

    C. Change the record owner

    D. Enable "View All" permission

    Answer: B

6. **What is the minimum access level a sharing rule can grant?**

    A. No access

    B. Read Only

    C. Read/Write

    D. Modify All

    Answer: B

7. **Which sharing mechanism is best for temporary, ad hoc record access?**

    A. Criteria-based sharing rule

    B. Owner-based sharing rule

    C. Manual sharing

    D. Public group

    Answer: C

8. **What happens to manual sharing when record ownership changes?**

    A. It stays intact.

    B. It is removed.

    C. It converts to a sharing rule.

    D. It is disabled but not deleted.

    Answer: B

9. **Why should public groups be used in sharing rules instead of adding users individually?**

    A. Groups improve dashboard visibility.

    B. Groups make rules more scalable and easier to manage.

    C. Groups increase record access speed.

    D. Groups reduce API calls.

    Answer: B

10. **In an organization with OWD set to Private, which option allows selective record sharing?**

    A. Page layout assignments

    B. Profiles

    C. Sharing rules

    D. Record types

    Answer: C

11. **Which sharing rule type would you choose if access needs are based on the record owner's role?**

    A. Owner-based sharing rule

    B. Criteria-based sharing rule

    C. Manual sharing

    D. Public group access

    Answer: A

12. **What is the main advantage of criteria-based sharing rules over owner-based rules?**

    A. They are easier to set up.

    B. They automatically adjust when field values change.

    C. They are faster in performance.

    D. They override OWD settings.

    Answer: B

13. **Which tool allows you to share a single record with a specific user?**

  A. Profile settings

  B. Manual sharing

  C. Role hierarchy

  D. Permission sets

  Answer: B

14. **Which is a valid access level option in a sharing rule?**

  A. Modify All

  B. View Setup and Configuration

  C. Read Only

  D. API Enabled

  Answer: C

15. **If OWD for Accounts is set to Public Read Only, what's the highest access a sharing rule can grant?**

  A. No Access

  B. Read Only

  C. Read/Write

  D. Modify All Data

  Answer: C

16. **What happens when too many complex sharing rules are created?**

  A. No impact on performance

  B. Increased record visibility

  C. Slower recalculation and performance degradation

  D. Automatic rule optimization

  Answer: C

17.   **Which sharing method is best for stable ownership patterns?**

A. Criteria-based

B. Owner-based

C. Manual sharing

D. Territory-based

Answer: B

18.   **Which feature should be reviewed first before creating any sharing rule?**

A. Page layouts

B. OWDs (Organization-Wide Defaults)

C. Validation rules

D. Field history tracking

Answer: B

19.   **What's a common use case for manual sharing?**

A. Permanent access between departments

B. Long-term cross-region collaboration

C. Short-term access for vacation coverage

D. Organization-wide visibility changes

Answer: C

20.   **What's a major limitation of manual sharing?**

A. Cannot be applied to custom objects

B. Requires Apex coding

C. Access is lost when record ownership changes

D. Only works in Classic UI

Answer: C

21. **Why might you avoid giving Read/Write access in a sharing rule?**

    A. It increases storage limits.

    B. It allows data editing, which may not be necessary.

    C. It hides field-level security.

    D. It removes profile permissions.

    Answer: B

22. **Which is NOT a valid reason to use sharing rules?**

    A. To grant cross-team record access

    B. To override OWD restrictions

    C. To assign page layouts

    D. To collaborate across departments

    Answer: C

23. **What's the best practice for documenting sharing rules?**

    A. Only document technical details

    B. Include business reasons, affected records, and related settings

    C. Store documentation outside Salesforce only

    D. Documentation is optional

    Answer: B

24. **When should you test sharing rules?**

    A. After production deployment only

    B. In sandbox before production deployment

    C. Never, testing is not required

    D. Only for criteria-based rules

    Answer: B

25. **Which team might use criteria-based rules for access?**

    A. HR team needing access to employee records with Status = Active

    B. Finance team wanting all invoices

    C. Admin team managing profiles

    D. IT team setting password policies

    Answer: A

26. **If a user's role changes, what happens to sharing rules that granted them access?**

    A. They always retain access.

    B. Access may be lost if based on previous role group membership.

    C. Access remains via manual sharing.

    D. It converts to OWD access.

    Answer: B

27. **Which component can be used within sharing rules to define the target audience?**

    A. Public groups

    B. Dashboards

    C. Page layouts

    D. Record types

    Answer: A

28. **Why is it risky to create too many criteria-based sharing rules?**

    A. They may ignore role hierarchy.

    B. They can cause performance issues with large data volumes.

    C. They require coding.

    D. They don't work with custom fields.

    Answer: B

29. **When should you use manual sharing instead of creating a sharing rule?**

    A. When access needs are permanent.

    B. When access applies to all records of a certain type.

    C. When access is for a single record temporarily.

    D. When configuring OWD.

    Answer: C

30. **Which statement about sharing rules is TRUE?**

    A. They override profile permissions.

    B. They only work with custom objects.

    C. They supplement, but do not replace, OWD settings.

    D. They apply only to users in the same role hierarchy.

    Answer: C

31. **What happens if you change the field value used in a criteria-based sharing rule?**

    A. Rule stops working permanently.

    B. Access updates automatically based on the new value.

    C. You must recreate the rule.

    D. It converts to an owner-based rule.

    Answer: B

32. **Which sharing model element is evaluated first before applying sharing rules?**

    A. Manual sharing

    B. OWDs (Organization-Wide Defaults)

    C. Criteria-based rules

    D. Apex sharing

    Answer: B

33. **Which scenario fits an owner-based sharing rule?**

    A. Share all cases with Priority = High

    B. Share opportunities owned by East Region Sales team

    C. Share accounts where Industry = Finance

    D. Share leads created in last 30 days

    Answer: B

34. **Why should you use public groups with sharing rules?**

    A. They speed up report generation.

    B. They make rule maintenance easier.

    C. They reduce license costs.

    D. They bypass OWD restrictions.

    Answer: B

35. **What is a common risk of over-sharing?**

    A. Reduced storage capacity

    B. Increased data security exposure

    C. Increased license costs

    D. Inaccurate dashboard totals

    Answer: B

36. **Which Salesforce feature recalculates sharing when rules change?**

    A. Data Loader

    B. Sharing recalculation engine

    C. Workflow queue

    D. Metadata API

    Answer: B

37. **When a record owner changes, manual sharing entries**

    A. Stay active indefinitely

    B. Are automatically removed

    C. Convert to criteria-based rules

    D. Become public read only

    Answer: B

38. **Why might you avoid chaining multiple sharing rules for one requirement?**

    A. It makes dashboards inaccurate.

    B. It can create unnecessary complexity and overlap.

    C. It increases storage costs.

    D. It hides audit trails.

    Answer: B

39. **If an object's OWD is Public Read/Write, the Sharing button**

    A. Always appears

    B. Is hidden because everyone already has full access

    C. Shows only for admins

    D. Works only in Classic

    Answer: B

40. **Which is a best practice for criteria-based rule fields?**

    A. Use fields that change frequently

    B. Use checkbox or picklist fields for simpler logic

    C. Use encrypted fields

    D. Use formula fields only

    Answer: B

41. **Which of these is NOT a characteristic of manual sharing?**

A. User-driven

B. Record-specific

C. Permanent for the record lifetime

D. Immediate effect

Answer: C

42. **Why is sandbox testing important before deploying sharing rules?**

A. To check if OWD will be disabled

B. To validate real user access scenarios without affecting production

C. To reduce license usage

D. To merge profiles automatically

Answer: B

43. **Which of these is an example of criteria-based sharing?**

A. Share accounts owned by Marketing team

B. Share all opportunities where Stage = Renewal

C. Share leads from East territory

D. Share all custom object records with one user

Answer: B

44. **What is the default behavior of role hierarchy?**

A. Users see only records they own.

B. Users higher in hierarchy can see records owned by subordinates.

C. Users in the same role see each other's records.

D. No automatic record access.

Answer: B

45. **In manual sharing, who can grant access to a record?**

   A. Any user in the org

   B. Record owner or users with Full Access

   C. Only system administrators

   D. Users in the same profile

   Answer: B

46. **What should you document for each sharing rule?**

   A. Only the rule name

   B. Rule logic, business justification, scope, and owner

   C. API field names only

   D. Nothing, documentation is optional

   Answer: B

47. **Which is a limitation of manual sharing in automation?**

   A. It cannot be applied in Lightning.

   B. It cannot be triggered by workflows or flows directly.

   C. It cannot be done for standard objects.

   D. It does not work with custom objects.

   Answer: B

48. **Which sharing method is most scalable for large groups?**

   A. Manual sharing

   B. Criteria-based sharing

   C. Owner-based sharing

   D. Public group sharing

   Answer: B

49. **When granting Read/Write access in a sharing rule, what risk is introduced?**

    A. Users may delete data.

    B. Users can modify records beyond intended scope.

    C. Profiles may change automatically.

    D. Reports may stop working.

    Answer: B

50. **What is the first step in designing an effective sharing rule?**

    A. Set OWD to Public Read/Write

    B. Map the collaboration requirement

    C. Create a public group

    D. Assign profiles

    Answer: B

51. **What's the main difference between owner-based and criteria-based rules?**

    A. Owner-based depends on record owner; criteria-based depends on field values.

    B. Owner-based is slower.

    C. Criteria-based ignores OWD.

    D. Owner-based works only for custom objects.

    Answer: A

52. **What should you do if a temporary exception keeps repeating?**

    A. Keep using manual sharing

    B. Create an appropriate sharing rule

    C. Delete the OWD

    D. Use profiles instead

    Answer: B

53. Which scenario requires updating sharing rules?

A. Territory realignment changes ownership patterns

B. Adding a new picklist value unrelated to access

C. Creating a report folder

D. Changing a dashboard owner

Answer: A

54. What's the role of public groups in sharing rules?

A. They help apply rules to multiple users without editing the rule itself.

B. They store criteria for field values.

C. They replace role hierarchy.

D. They create OWDs.

Answer: A

55. What should be done quarterly for sharing models?

A. Rebuild the role hierarchy

B. Review and validate that rules still meet business needs

C. Delete all manual sharing

D. Reset all passwords

Answer: B

56. Why might you avoid overly complex criteria in rules?

A. It makes UI slow.

B. It slows down sharing recalculations.

C. It increases license cost.

D. It disables role hierarchy.

Answer: B

57. **Which sharing approach should be used for a one-time partner account review?**

    A. Manual sharing

    B. Owner-based rule

    C. Criteria-based rule

    D. OWD

    Answer: A

58. **In a merger, what's the first step before redesigning the sharing model?**

    A. Delete old rules

    B. Map new teams' data needs and compare with existing rules

    C. Increase storage limits

    D. Change all OWD to Public Read Only

    Answer: B

59. **Which of the following is NOT a factor in choosing between sharing rule types?**

    A. Stability of ownership patterns

    B. Field value changes

    C. Record type assignments

    D. Number of custom objects in org

    Answer: D

60. **Which is the safest principle for sharing model design?**

    A. Share as much as possible for collaboration

    B. Start simple and expand only as required

    C. Give Read/Write to all users

    D. Avoid using public groups

    Answer: B

CHAPTER 5

Advanced Techniques—Apex-Managed Sharing and Beyond

Apex-managed sharing is not about replacing Salesforce's standard sharing model. It is about extending it—carefully, deliberately, and only when the business reality demands it. It is the mechanism Salesforce provides for expressing access logic that depends on *context*, not just configuration.

In the earlier chapters of this book, we explored Salesforce's record visibility model from its foundational layers upward. Organization-Wide Defaults established the baseline of trust. Role hierarchies defined vertical visibility. Sharing rules and manual sharing added flexibility, allowing access to flow across teams and departments without dismantling the core security structure.

This chapter builds on everything you have learned so far and moves into the most advanced layer of Salesforce sharing design. Thus, you'll find it more complex and architecturally nuanced, where a strategic perspective adds deeper value and synthesis.

You will explore

- How share objects work and how Salesforce evaluates them.

- How to design meaningful and auditable Row Causes.

- How to think about granting and revoking access as a lifecycle.

- Architectural patterns that keep custom sharing logic maintainable.

- Real-world scenarios where Apex-managed sharing is the only viable option.

© Sandhya Sharma 2026

S. Sharma, *Unlocking Salesforce Data*, https://doi.org/10.1007/979-8-8688-2305-3_5

- Performance implications and strategies for scaling safely.

- Governance practices that prevent custom sharing from becoming technical debt.

You will also engage with structured brainstorming exercises designed to help you recognize when Apex-managed sharing is appropriate—and when it is not. By the end of this chapter, you should not only understand *how* Apex-managed sharing works, but *when* to use it, *why* it exists, and *how* to design it responsibly.

When Declarative Sharing Reaches Its Limits

For most organizations, Salesforce's layered model is more than sufficient. In fact, many successful Salesforce implementations operate entirely within declarative sharing for years. These tools are predictable, transparent, and relatively easy to maintain. They encourage thoughtful design and discourage over-engineering.

And yet, as organizations mature, their processes rarely remain simple.

Growth introduces nuance. Mergers introduce exceptions. Compliance introduces constraints. Temporary teams, external collaborators, multi-dimensional reporting lines, and cross-object dependencies introduce requirements that do not fit neatly into static rules. At this stage, architects often find themselves trying to stretch declarative sharing beyond its natural boundaries—adding more roles, more public groups, more criteria, and more documentation to explain why something works the way it does.

This chapter exists for that moment.

From Static Rules to Context-Aware Visibility

Declarative sharing operates on a simple premise: if a record meets a condition, access is granted. That condition might be ownership, a field value, or a role relationship. The strength of this approach is its consistency. The weakness is that it assumes access requirements are stable and easily described.

Many real-world scenarios violate this assumption.

Consider access that depends on

- The relationship between multiple records, not just one

- A user's temporary assignment or certification

- A project's lifecycle stage combined with external funding constraints

- Regulatory obligations that vary by geography or donor

In such cases, access is not merely a question of *what the record is*, but *what is happening around it*. Declarative tools can describe static states very well. They struggle with dynamic situations.

Apex-managed sharing introduces the ability to evaluate these situations programmatically and express the outcome directly in Salesforce's underlying sharing tables. Instead of relying on Salesforce to infer access from configuration, the system is told explicitly: *this user should see this record, for this reason, right now.*

This is a powerful capability—and one that must be treated with respect.

What Apex-Managed Sharing Really Is

At its core, Apex-managed sharing is a way to interact with Salesforce's internal sharing mechanism. Every object that supports sharing—standard or custom—has an associated "share object" that stores record-level access entries. These entries answer three fundamental questions:

- Which record is being shared?

- Who is receiving access?

- What level of access is being granted, and why?

When declarative sharing rules run, Salesforce populates these tables automatically. Manual sharing adds rows when a user clicks a button. Apex-managed sharing allows developers to create, modify, or remove these rows directly, based on logic that Salesforce cannot express declaratively.

This does **not** bypass Salesforce security. Profiles, permission sets, field-level security, and object permissions still apply. Apex-managed sharing only determines *which records are visible*, not *what a user can do with them*.

This distinction is critical. Apex-managed sharing operates within Salesforce's security framework—it does not override it.

Why Apex-Managed Sharing Matters to Architects

For administrators, Apex-managed sharing may appear intimidating. It introduces code, testing, deployment cycles, and long-term maintenance considerations. For architects, however, it represents something else entirely: control.

Control over

- When access is granted.

- When access is revoked.

- Why access exists.

- How access scales as data volume grows.

Without Apex-managed sharing, complex access models often devolve into compromises. Organizations either

- Over-share data "to be safe," increasing compliance risk, or

- Under-share data, forcing users into workarounds and
  shadow systems

Neither outcome is acceptable in mature implementations.

Apex-managed sharing allows architects to encode business intent directly into the access model. It makes the sharing logic explicit, testable, and auditable. When implemented well, it reduces ambiguity rather than increasing it.

When Apex-Managed Sharing Becomes Necessary

It is important to be clear: Apex-managed sharing should never be the default choice. It is a solution for specific classes of problems—problems that cannot be solved cleanly with standard tools.

Common signals that declarative sharing has reached its limit include the following:

- **Access Depends on Related Records:**

  For example, a project should only be visible if its related funding agreement is external, *and* the project has moved beyond an internal review stage. Sharing rules cannot evaluate conditions across multiple objects in this way.

- **Access Must Change Automatically Over Time:**

  Temporary teams, contractors, auditors, or consultants may require access for a defined window. Relying on manual cleanup is risky and error-prone.

- **Access Depends on Attributes of the User, Not Just Their Role:**

  Certifications, skills, or participation in temporary initiatives often cut across formal hierarchies.

- **Access Logic Must Be Reversible and Auditable:**

  In regulated environments, it is not enough to grant access. You must also be able to explain why it was granted and demonstrate that it was removed when conditions changed.

When these requirements appear, Apex-managed sharing is often the cleanest—and safest—solution.

A Realistic Example: Disaster Response in a Nonprofit Org

Consider a nonprofit organization coordinating disaster relief efforts across multiple regions.

When a new relief project is created:

- Country managers in the affected region must immediately see and update the project.

- A global risk committee must gain visibility if the project is classified as high risk.

- Donor representatives must see only the projects they fund, and only in read-only mode.

These requirements depend on

- Geography

- Risk classification

- Funding relationships

- User affiliation

No single sharing rule can express this logic. Attempting to do so would result in a maze of public groups, duplicated rules, and brittle criteria. Apex-managed sharing allows the organization to evaluate each condition and grant access explicitly, with a clear reason attached to each decision.

Just as importantly, when the project closes or funding ends, access can be revoked automatically—without relying on memory, manual cleanup, or hope.

Lessons Learned from the Field

Organizations that succeed with Apex-managed sharing tend to share several characteristics.

They are deliberate. They resist the temptation to use code where configuration would suffice. Apex-managed sharing is introduced only after simpler options are exhausted.

They are explicit. Every share entry has a clearly defined reason. This makes audits, debugging, and future enhancements far easier.

They design for removal, not just addition. Access is treated as temporary by default, even when it may last for years.

They think in volume. Solutions are designed to work at scale, not just in development sandboxes.

And perhaps most importantly, they treat sharing logic as architecture—not as an afterthought.

Share Object Concepts—Understanding Salesforce's Hidden Visibility Layer

Before an architect can design or reason about Apex-managed sharing, it is essential to understand where Salesforce actually stores record-level access decisions. While administrators interact with sharing through setup screens, buttons, and rules, the platform itself relies on a set of internal data structures—commonly referred to as *share objects*—to determine who can see what.

These objects are rarely discussed in depth outside advanced architecture conversations, yet they form the backbone of Salesforce's runtime security model. Every query, report, API call, and UI render ultimately checks these tables before returning records to a user.

Understanding share objects is not about memorizing field names. It is about understanding how Salesforce *thinks* about visibility.

Share Objects As the System of Record for Access

Salesforce does not dynamically calculate sharing logic every time a user opens a record. Instead, it materializes access decisions into share objects and consults them when needed. This design choice is deliberate. It allows the platform to scale to millions—or billions—of records without recalculating permissions on the fly.

For every object that supports sharing, Salesforce maintains a corresponding share object. The naming convention is consistent:

- Account → AccountShare

- Opportunity → OpportunityShare

- Case → CaseShare

- CustomObject__c → CustomObject__Share

These share objects are not visible to most users, but they are first-class citizens in the data model. Each row represents a single, explicit access grant between a record and a user or group.

In effect, share objects are the *ledger* of record visibility.

What a Share Record Represents?

Each row in a share object answers a very specific question:

"Why does this user have this level of access to this record?"

The answer is expressed through a small set of fields that appear simple but carry significant architectural meaning.

At a conceptual level, every share record contains four critical pieces of information:

1. **Which record is being shared?**

2. **Who is receiving access?**

3. **What level of access is granted?**

4. **Why does that access exist?**

These dimensions are what make Apex-managed sharing possible—and auditable.

ParentId—The Record Being Shared

The ParentId field points to the record whose visibility is being extended. This is always the Id of the object's main record, such as an Account, Opportunity, or custom object record.

From an architectural perspective, ParentId establishes the scope of the sharing decision. Apex-managed sharing is always record-specific. Even when access is granted to hundreds of users, each access decision is stored as an individual relationship to a specific record.

This design has two important implications:

- Sharing is precise. You can reason about access one record at a time.

- Sharing can grow large quickly. High-volume objects can accumulate millions of share rows if not designed carefully.

Architects must keep both precision and scale in mind when working with share objects.

UserOrGroupId—Who Receives Access

The UserOrGroupId field identifies the recipient of access. This can reference

- An individual user

- A public group

- A role or role hierarchy node

- A territory (where applicable)

From a design standpoint, this is where many long-term maintainability decisions are made. Granting access to individual users may solve an immediate problem, but it rarely scales well. Granting access to groups or roles allows access models to evolve without rewriting sharing logic.

Experienced architects almost always treat UserOrGroupId as an abstraction layer. Instead of asking, "Which users need access?", they ask, "Which group represents this access requirement?"

That distinction becomes critical as organizations grow.

AccessLevel—How Much Access Is Granted

AccessLevel defines the degree of visibility granted to the recipient. For most objects, this is limited to

- **Read Only**

- **Read/Write**

Importantly, AccessLevel does not override object permissions, field-level security, or profile restrictions. It simply defines the *maximum* access the user may have to the record—subject to all other security layers.

This is a frequent point of confusion for teams new to Apex-managed sharing. Granting Edit access does not guarantee that a user can edit the record; it only makes editing possible if their profile and permission sets allow it.

From an architectural perspective, AccessLevel should be treated conservatively. Grant the minimum level necessary to achieve the business outcome. Over-granting access increases risk and complicates audits.

RowCause—Why the Access Exists

RowCause is the most underappreciated—and most important—field in a share object.

It explains *why* the share row exists.

Salesforce uses RowCause values to distinguish between different sources of access, such as

- Ownership

- Sharing rules

- Manual sharing

- Programmatic sharing

In Apex-managed sharing, RowCause becomes a critical governance tool. Custom RowCause values allow architects to encode business intent directly into the data model. Instead of a generic "this user has access," the system can express "this user has access because they are part of the disaster response team" or "because they fund this project."

This distinction matters enormously during audits, debugging, and future redesigns.

Without meaningful RowCause values, share objects become opaque. With them, they become self-documenting.

How Salesforce Evaluates Share Objects at Runtime

When a user attempts to access a record, Salesforce evaluates visibility in layers. While the exact internal order is abstracted away, conceptually the platform checks

- Object-level access (profiles and permission sets)

- Organization-Wide Defaults

- Role hierarchy access

- Explicit sharing entries (from share objects)

- Manual and Apex-managed sharing rows

If any path grants access, the record becomes visible—subject to the most restrictive access level across all applicable paths.

This means that Apex-managed sharing does not replace other sharing mechanisms. It supplements them. It creates additional access paths that Salesforce evaluates alongside declarative ones.

Architects should always think in terms of *access paths*, not isolated rules.

Declarative Sharing vs. Apex-Managed Sharing in the Share Table

One of the most important conceptual shifts when moving to Apex-managed sharing is understanding that declarative and programmatic sharing coexist in the same physical tables.

A single record might have the following:

- One share row created by a role hierarchy

- Another created by a criteria-based sharing rule

- A third created by Apex logic

- A fourth created manually by a user

Salesforce does not distinguish between these at runtime. All are simply rows in the share object. The only way to tell them apart is RowCause.

This reinforces why disciplined RowCause governance is non-negotiable. Without it, you lose the ability to explain *why* access exists—or how to safely remove it.

Share Objects and Custom Objects

Custom objects behave no differently than standard ones when it comes to sharing. If a custom object's Organization-Wide Default is set to Private or Public Read Only, Salesforce generates a corresponding share object.

This is a powerful feature. It means architects can design entirely custom data models and still rely on Salesforce's native sharing infrastructure. There is no separate security engine for custom objects.

However, this also means that poorly designed custom sharing logic can have platform-wide consequences. Custom objects often store high-volume transactional data. Apex-managed sharing on these objects must be designed with extreme care.

Share Object Volume and Data Growth

One of the most common mistakes teams make is underestimating how quickly share objects grow.

A single record shared with

- Ten users creates ten share rows

- Ten groups creates ten share rows

- Ten roles across five regions creates dozens more

Multiply this by thousands—or millions—of records, and the volume becomes significant.

Share objects count toward data storage. They also impact query performance and sharing recalculation times. Architects must always consider

- How many records will exist?

- How many recipients will receive access?

- How often will access change?

Designing Apex-managed sharing without answering these questions is irresponsible.

Share Objects As a Design Constraint, Not an Implementation Detail

It is tempting to treat share objects as an internal detail—something developers interact with but architects need not worry about. This is a mistake.

Share objects are a design constraint. They enforce

- One row per access decision

- Explicit storage of visibility

- Clear separation between access grant and access reason

When architects embrace this constraint, better designs emerge. Access becomes intentional. Revocation becomes first-class. Auditing becomes feasible.

When architects ignore it, sharing logic becomes fragile and opaque.

A Mental Model for Architects

A useful way to think about share objects is this:

- Profiles define *what* a user can do.

- Permission sets refine *how* they can do it.

- Share objects define *which records* they can see.

Apex-managed sharing operates exclusively in the third dimension.

It does not grant new powers. It does not bypass security. It simply tells Salesforce, "This record should be visible through this explicit relationship."

Once you internalize this mental model, the rest of Apex-managed sharing becomes far easier to reason about.

Looking Ahead

With a solid understanding of share object concepts, we can now move into one of the most critical—and most frequently neglected—topics in advanced sharing design: **Row Cause Governance**.

This is where Apex-managed sharing either becomes a sustainable architectural asset or a long-term liability.

Row Cause Governance—Turning Sharing Logic into an Auditable System

If share objects are the engine of record visibility, then *RowCause* is the steering wheel.

It is the field that explains intent. It tells future administrators, auditors, and architects *why* access exists—not just that it does. In Apex-managed sharing, RowCause is not a technical afterthought; it is a governance contract between the system and the people who maintain it.

Many Salesforce orgs implement programmatic sharing successfully in the short term, only to struggle years later when no one can explain where access is coming from. Almost without exception, the root cause is poor RowCause governance.

Why Row Cause Matters More Than You Think

At a technical level, RowCause is simply a value stored on a share record. At an architectural level, it is the only reliable way to distinguish one access path from another.

Salesforce does not maintain separate tables for "manual sharing," "rule-based sharing," and "Apex sharing." Everything lives together. When a record is visible to a user, Salesforce does not tell you which mechanism granted access—it only confirms that access exists.

RowCause is what restores meaning to that visibility.

Without disciplined RowCause usage, every share row looks the same. With it, the sharing model becomes explainable.

Standard Row Causes and Their Limitations

Salesforce provides a small set of standard RowCause values out of the box. These typically represent broad system-level reasons such as ownership, rules, or manual actions.

These standard causes are useful, but they are intentionally generic. They answer questions like

- "Is this access automatic or manual?"

- "Did it come from configuration or code?"

They do *not* answer

- "Which business rule granted this access?"

- "Which program or process required it?"

- "Is this access still valid today?"

As soon as Apex-managed sharing enters the picture, standard RowCause values become insufficient. Custom logic demands custom explanations.

Custom Row Causes As Business Documentation

Custom RowCause values allow architects to encode business meaning directly into the data layer.

Instead of treating sharing as a technical artifact, RowCause allows it to become part of the organization's governance narrative. Each custom value should correspond to a clearly defined access policy.

For example,

- Access granted because a user is part of a regional disaster response team.

- Access granted due to a funding relationship.

- Access granted for compliance review.

- Access granted for time-bound collaboration.

The exact labels are less important than the principle: **each RowCause should map to one—and only one—business justification.**

This one-to-one mapping is what enables safe evolution of the sharing model over time.

Designing a Row Cause Taxonomy

One of the most effective practices in mature Salesforce orgs is treating RowCause values as a taxonomy rather than a collection.

Instead of inventing RowCause values reactively, architects define them intentionally. Each value

- Has a clear name

- Has a documented purpose

- Has a known lifecycle

- Has an owning team

A simple taxonomy might group RowCause values by domain

- Program-based access

- Compliance-based access

- Operational access

- Temporary access

This structure makes it immediately obvious why access exists and who is responsible for it.

Row Cause and Revocation Strategy

One of the greatest strengths of Apex-managed sharing is reversibility—but only if RowCause is used correctly.

When access conditions change, the system must be able to identify *which* share rows to remove. If RowCause values are ambiguous or reused across unrelated scenarios, revocation becomes dangerous.

Teams may hesitate to remove access because they cannot be certain what will break.

Clear RowCause governance eliminates this fear. When each share row carries an explicit reason, revocation logic becomes precise rather than speculative.

This is where many early Apex-managed sharing implementations fail. They focus heavily on granting access and treat removal as an afterthought. Years later, the org is littered with orphaned access that no one dares to touch.

Auditing and Compliance Implications

From an audit perspective, RowCause is invaluable.

When auditors ask, "Why does this user have access to this record?", a well-governed sharing model can answer confidently. Each share row points to a documented business rule.

Without RowCause discipline, audits become manual, time-consuming, and stressful. Teams resort to screenshots, tribal knowledge, and best guesses.

In regulated industries—nonprofits handling donor data, healthcare organizations managing sensitive cases, or financial institutions dealing with client records—this difference is not academic. It can determine whether an audit passes or fails.

Row Cause Lifecycle Management

RowCause values should not be permanent by default.

Every access policy has a lifecycle:

- It is introduced to meet a requirement.

- It remains active while that requirement exists.

- It should be retired when the requirement ends.

Mature organizations treat RowCause values the same way they treat permission sets or integration endpoints: they review them periodically.

Questions worth asking during reviews:

- Is this RowCause still relevant?

- Does it still align with current business processes?

- Are there share rows using this cause that should be removed?

- Has the access policy changed in scope or intent?

Ignoring these questions leads to access creep—the slow accumulation of unnecessary visibility over time.

Naming Conventions That Scale

RowCause names should be readable by humans, not just developers.

Avoid cryptic abbreviations or overly technical labels. A future admin should be able to understand the intent without opening code or documentation.

Good naming practices are as follows:

- Use business language

- Avoid references to implementation details

- Prefer descriptive over concise

A RowCause named after a business policy will outlive one named after a project or sprint.

Row Cause and Cross-Team Collaboration

Common Anti-Patterns to Avoid

Several RowCause anti-patterns appear repeatedly in struggling orgs:

- **Reusing a Single Custom RowCause for Everything:**

  This defeats the purpose entirely. It makes all programmatic access indistinguishable.

- **Naming RowCauses After Developers or Tickets:**

  These labels lose meaning quickly and provide no business context.

- **Failing to Document RowCause Intent:**

  Even well-named values need lightweight documentation.

- **Treating RowCause As Optional:**

  In Apex-managed sharing, it is not optional—it is foundational.

Recognizing and avoiding these patterns early prevents years of cleanup work later.

Row Cause As an Architectural Contract

Ultimately, RowCause is a contract.

It is a promise that every access grant has a reason, and that reason is explicit, traceable, and reversible. It signals architectural maturity.

Organizations that respect this contract find that Apex-managed sharing becomes a powerful, trusted tool. Organizations that ignore it often come to see programmatic sharing as dangerous or unmanageable—not because it is, but because it was never governed.

With RowCause governance in place, Apex-managed sharing becomes predictable rather than mysterious. Architects can now design access flows that are

- Intentional

- Auditable

- Reversible

- Scalable

This foundation allows us to move confidently into the next topic: **Access and Revocation Patterns**—where we examine how visibility is granted, monitored, and safely withdrawn over time.

Access and Revocation—Designing Visibility As a Lifecycle, Not an Event

One of the most common mistakes teams make when implementing Apex-managed sharing is treating access as a one-time decision. A record becomes visible, the requirement is met, and the work is considered done.

In reality, access is never static.

People change roles. Projects evolve. Funding ends. Regulations shift. The moment access is granted, the question of *when it should be removed* already exists—even if no one asks it yet.

Apex-managed sharing is powerful precisely because it allows architects to think in terms of **access lifecycles** rather than permanent permissions. This section explores how to design sharing logic that anticipates change, enforces discipline, and avoids the slow accumulation of unnecessary visibility.

Access Is a Temporary State, Even When It Feels Permanent

Declarative sharing rules often feel permanent because they are tied to stable constructs like roles or record ownership. Apex-managed sharing, by contrast, tends to be introduced for conditions that are fluid by nature.

Examples include the following:

- A project entering a specific phase

- A user joining a temporary task force

- A compliance review window opening

- A contractual relationship becoming active

In each of these cases, access exists because a condition is true *right now*. That condition may not remain true indefinitely.

Architects who recognize this early design sharing logic that treats every grant as provisional—even if the duration is long.

The Symmetry Between Granting and Revoking Access

In well-architected systems, granting and revoking access are symmetrical processes. They are two sides of the same rule.

Unfortunately, many implementations focus almost entirely on the "grant" side. Access is added when criteria are met, but little thought is given to what happens when criteria are no longer valid.

This asymmetry leads to what is often called *access residue*: visibility that lingers long after its justification has disappeared.

A disciplined Apex-managed sharing design always answers two questions together:

1. Under what conditions should access be granted?

2. Under what conditions should that same access be removed?

If the second question cannot be answered clearly, the first should not be implemented yet.

Access Driven by State Changes

One of the most common patterns for Apex-managed sharing involves state transitions.

A record moves from one phase to another, and visibility changes accordingly. This might be a project advancing through stages, a case escalating in severity, or a record entering a review cycle.

What matters is not the absolute state, but the *transition*. Access often needs to change when something crosses a boundary.

Designing for transitions rather than static states ensures that revocation is not forgotten. When access logic explicitly accounts for "before" and "after," cleanup becomes part of the design instead of an afterthought.

Time-Bound Access As a First-Class Concept

Some access requirements are explicitly time-bound.

Contractors may need access for the duration of their engagement. External reviewers may require visibility during an audit window. Temporary task forces may dissolve after a milestone is reached.

In these cases, Apex-managed sharing is often the only viable approach, because declarative sharing cannot reason about time without manual intervention.

Time-bound access introduces an important architectural consideration: **revocation must be scheduled, not reactive**.

Rather than waiting for a human to notice that access should be removed, mature systems build revocation into their timelines. The system knows when access expires, and removal is treated as a predictable event.

This approach dramatically reduces risk, particularly in regulated environments where overstaying access can have legal consequences.

Revocation Is More Than Deletion

It is tempting to think of revocation as simply removing a share row. From a technical perspective, that is accurate. From an architectural perspective, it is incomplete.

Revocation should be observable.

When access is removed, teams should be able to answer:

- What access was removed?

- Why was it removed?

- When did it happen?

- Was it expected or exceptional?

Without visibility into revocation events, teams lose confidence in the sharing model. Access disappears, users complain, and administrators scramble to reverse changes without understanding root causes.

A well-designed Apex-managed sharing strategy treats revocation as a documented event, not a silent side effect.

The Risk of Orphaned Access

One of the most serious long-term risks in Apex-managed sharing is orphaned access—share rows that no longer correspond to any active business rule.

These often arise when

- Projects are archived but access remains.

- Users change roles but retain old visibility.

- Processes are redesigned without cleaning up legacy logic.

Orphaned access is particularly dangerous because it is invisible. Users may not even realize they have access they should not have, and administrators may not know where to look.

Preventing orphaned access requires two things:

- Clear RowCause governance

- Regular review of active share rows

Neither can be substituted for the other.

Access Reviews As a Governance Practice

In mature organizations, access reviews are not limited to profiles and permission sets. Programmatic sharing is included as a first-class citizen in governance discussions.

Periodic reviews ask questions such as

- Which RowCause values are currently active?

- How many share rows exist under each cause?

- Are the underlying business justifications still valid?

- Are there share rows associated with inactive users or archived records?

These reviews do not need to be frequent, but they must be intentional. Even an annual review can dramatically reduce long-term risk.

Designing for Predictable Revocation

One hallmark of good Apex-managed sharing design is predictability.

Users should not experience access changes as random or mysterious. When access is revoked, it should align with recognizable business events:

- A project stage changes

- A contract ends

- A user leaves a team

- A review period closes

When revocation is aligned with business milestones, it is easier to explain, defend, and troubleshoot.

Unpredictable revocation erodes trust in the system. Predictable revocation reinforces it.

Avoiding Overlapping Access Paths

Another challenge in access management is overlapping access paths.

A user might gain visibility through the following:

- A sharing rule

- Apex-managed sharing

- Manual sharing

- Ownership changes

When multiple mechanisms grant the same access, revocation becomes complicated. Removing one path may have no visible effect because another still exists.

This is not inherently wrong, but it must be intentional. Architects should know which access paths are expected to overlap and which are meant to be exclusive.

Clear documentation and disciplined design prevent the confusion that arises when access cannot be traced to a single cause.

Revocation in Matrix Organizations

Matrix organizations pose a particular challenge for access revocation.

Users may participate in multiple teams simultaneously, each with its own access requirements. Visibility might depend on a combination of skills, certifications, regions, or assignments.

In such environments, Apex-managed sharing often reflects *membership* rather than hierarchy. Access exists because the user belongs to a set, not because of where they sit in the org chart.

Revocation, therefore, must respond to membership changes. When someone leaves a team, access should adjust automatically.

Failing to design for this leads to former team members retaining visibility indefinitely—an especially common issue in project-driven organizations.

Handling Exceptions Without Breaking the Model

No sharing model is perfect. There will always be exceptions.

A senior leader may need temporary visibility. A regulator may require access outside normal rules. A crisis may demand rapid collaboration.

The temptation is to hard-code these exceptions into the sharing logic. This is rarely a good idea.

Exceptions should remain exceptional. Apex-managed sharing supports this by allowing targeted, well-documented grants that can be explicitly revoked later.

The key is ensuring that exceptions do not silently become permanent features.

The Cost of Ignoring Revocation

Organizations that neglect revocation eventually pay the price.

Symptoms include the following:

- Users seeing records they no longer recognize

- Compliance teams raising concerns

- Administrators afraid to touch sharing logic

- Architects inheriting systems no one fully understands

At that point, cleaning up access is far more expensive than designing it properly from the start.

Revocation discipline is not an optional refinement—it is the difference between a sustainable sharing model and a fragile one.

Designing for the Long Term

Access and revocation design is ultimately about humility.

Architects must accept that today's requirements will change. The system must be resilient to that change without requiring heroics.

By treating access as a lifecycle, designing revocation alongside grants, and aligning visibility changes with business events, Apex-managed sharing becomes a reliable tool rather than a liability.

With these principles in place, we can now zoom out and examine how these patterns fit into a broader **architectural design strategy**—one that balances flexibility, performance, and governance at scale.

Architectural Design—Building Apex-Managed Sharing That Survives Change

Apex-managed sharing should never exist as an isolated technical solution. When introduced without architectural intent, it quickly becomes one of the most fragile parts of a Salesforce org—difficult to understand, risky to modify, and expensive to maintain.

Strong architectural design turns Apex-managed sharing into the opposite: a controlled extension of the standard sharing model that behaves predictably, scales gracefully, and remains intelligible years after it was written.

This section explores how to design Apex-managed sharing as a **system**, not a shortcut.

Apex-Managed Sharing As an Extension, Not a Replacement

One of the most important architectural decisions is philosophical rather than technical: Apex-managed sharing is not meant to replace Salesforce's declarative sharing model.

Role hierarchies, Organization-Wide Defaults, and sharing rules should continue to do as much work as they reasonably can. Apex-managed sharing exists to handle the *edges*—the scenarios that fall outside what declarative tools can express.

Architects who treat Apex-managed sharing as a replacement for standard sharing often reintroduce complexity that Salesforce already addresses. Those who treat it as an extension keep their systems simpler and more resilient.

A useful mental model is this:

- Declarative sharing defines the **baseline.**

- Apex-managed sharing handles **exceptions and nuance.**

- Manual sharing addresses **human judgment.**

When these responsibilities are clearly separated, the system remains comprehensible.

Centralizing Sharing Logic

A recurring failure pattern in Apex-managed sharing is fragmentation. Sharing logic appears in multiple triggers, flows, batch jobs, and integrations, each adding or removing access based on local assumptions.

Over time, no single place reflects the full access model.

Architectural discipline demands centralization—not necessarily in a single file, but in a single *conceptual layer*. All programmatic sharing decisions should be routed through a known, documented boundary.

This approach offers several benefits:

- Changes to access logic are easier to reason about.

- Audits can trace decisions to a defined source.

- Performance tuning is localized rather than scattered.

Centralization does not mean rigidity; it means clarity.

Designing for Business Language, Not Technical Constructs

Sharing models are ultimately business artifacts. They reflect how an organization thinks about trust, responsibility, and collaboration.

Architectural design should therefore align sharing logic with business language rather than technical jargon.

Instead of thinking in terms of "insert share rows when field X equals Y," think in terms of

- "Who should be able to see this, and why?"

- "What business event justifies this access?"

- "What would make this access no longer appropriate?"

When the architecture mirrors business reasoning, it becomes easier to explain, document, and govern.

Layering the Sharing Model

Mature Apex-managed sharing architectures are layered.

At the bottom is the **data layer**, where share objects exist and access is technically enforced. Above that sits the **decision layer**, where logic determines whether access should exist. At the top is the **policy layer**, which expresses business intent and governance rules.

Separating these layers conceptually—even if they live close together technically— prevents accidental coupling between business rules and low-level mechanics.

It also makes it easier to evolve one layer without destabilizing the others.

Designing for Change, Not Stability

A paradox of sharing architecture is that the most stable systems are designed with change in mind.

Organizations evolve. Teams reorganize. Regulations tighten. What seems permanent today often turns out to be transitional.

Architectural design must assume that

- New access rules will be added.

- Old rules will be retired.

- Existing rules will be refined.

Apex-managed sharing should therefore be designed to absorb change incrementally. Adding a new access rule should not require rewriting existing ones. Removing a rule should not cause unintended side effects.

This is achieved not through cleverness, but through restraint.

Avoiding Implicit Dependencies

One subtle architectural risk is implicit dependency—where sharing logic quietly relies on assumptions that are not documented or enforced.

For example:

- Assuming a field will never change its meaning.

- Assuming a role structure will remain stable.

- Assuming a process always runs in a certain order.

These assumptions may be true today, but architecture must survive tomorrow.

Good design surfaces dependencies explicitly. It acknowledges where sharing logic relies on upstream processes and documents those relationships clearly.

When assumptions are visible, they can be challenged, validated, or revised.

Sharing Architecture in Large Orgs

As Salesforce orgs grow, sharing architecture becomes one of the most performance-sensitive areas of the system.

Large orgs often face

- High record volumes

- Frequent data changes

- Multiple concurrent sharing mechanisms

Architectural design must therefore consider not just correctness, but **impact**.

Apex-managed sharing should minimize unnecessary recalculation and avoid cascading effects. Changes to one area of the system should not trigger massive, unintended sharing updates elsewhere.

This requires thinking in terms of scope:

- Which records are affected?

- Which users are affected?

- How often will this logic run?

Architects who ask these questions early avoid painful refactors later.

Designing for Transparency and Debuggability

Apex-managed sharing failures are notoriously hard to diagnose if architecture does not prioritize transparency.

When users report unexpected access—or lack of access—teams must be able to answer

- Which mechanism granted or removed access?

- When did it happen?

- Under what conditions?

Architectural design should make these questions answerable without forensic investigation.

Clear RowCause usage, consistent naming, and documented access paths all contribute to a system that can be debugged under pressure.

Transparency is not a luxury; it is an operational necessity.

Supporting Multiple Access Patterns Without Chaos

Complex organizations often require multiple access patterns to coexist:

- Regional access

- Functional access

- Temporary access

- Compliance-driven access

Architectural design must allow these patterns to coexist without interfering with one another.

This is achieved by

- Keeping access rules narrowly scoped

- Avoiding overlapping responsibilities

- Clearly defining precedence and interaction

When patterns are cleanly separated, complexity remains manageable even as the system grows.

Governance As an Architectural Concern

Governance is often treated as an afterthought—something imposed once a system is already live. In Apex-managed sharing, governance must be baked into the architecture from the start.

This includes

- Naming conventions

- Documentation standards

- Review cycles

- Ownership of access rules

Architectural design should make governance easy to apply rather than difficult to retrofit.

When governance aligns with architecture, compliance becomes routine instead of disruptive.

Designing for People, Not Just Systems

Finally, good sharing architecture recognizes that humans—not just systems—interact with access models.

Administrators inherit systems they did not design. Auditors examine logic they did not build. Developers maintain code they did not write.

Architectural clarity is a form of empathy. It acknowledges that future readers deserve to understand why things exist, not just how they work.

Apex-managed sharing that respects this principle remains maintainable long after its original authors have moved on.

A Foundation for the Next Layer

Architectural design creates the foundation on which real-world scenarios, performance strategies, and governance practices are built.

With a solid architecture in place, Apex-managed sharing becomes a predictable, auditable, and scalable part of the Salesforce security model—rather than a source of uncertainty.

In the next section, we will examine **real-world scenarios** where these architectural principles come to life, illustrating how organizations apply Apex-managed sharing to solve problems that cannot be addressed through declarative means alone.

Real-World Scenarios—When Apex-Managed Sharing Becomes Essential

Apex-managed sharing rarely enters a Salesforce architecture by design alone. More often, it appears after repeated attempts to solve a problem with standard tools fail quietly or awkwardly. These scenarios share a common theme: access decisions depend on **context**, not just ownership, hierarchy, or simple field values.

This section explores real-world patterns where Apex-managed sharing is not merely convenient, but necessary. Each scenario highlights why declarative sharing falls short and how programmatic control enables precision without sacrificing governance.

Scenario 1: Access Driven by Related Records Across Multiple Objects

One of the most common limitations of sharing rules is their inability to evaluate conditions across related objects. While formulas can reference parent fields in some contexts, sharing rules themselves remain constrained to the record being shared.

Consider a nonprofit managing long-term community development initiatives. A *Project* record represents the initiative, but access depends on information stored elsewhere:

- Funding agreements determine who is financially accountable.

- Donor profiles specify reporting rights.

- Regional oversight teams vary by geography and risk classification.

Access to the Project should change when

- A funding agreement is amended.

- A donor withdraws or joins.

- The project's risk rating changes based on assessments stored on a related object.

Attempting to model this with declarative sharing quickly breaks down. You either over-share by granting access broadly or under-share by locking out users who legitimately need visibility.

Apex-managed sharing allows the system to evaluate the *entire relationship graph*— Project, Funding Agreement, Donor, Region—and grant access accordingly. Visibility becomes a result of relationships, not static rules.

Scenario 2: Time-Bound Access with Automatic Expiry

Time-based access is another area where declarative tools struggle. While manual sharing can grant temporary access, it relies on humans to remember to revoke it.

In many organizations, access must expire automatically:

- External auditors reviewing records for a fixed period

- Consultants embedded for a defined engagement

- Contractors assigned to short-term initiatives

In one global NGO, contractors supported emergency response efforts for specific disasters. Each contractor needed access only while their assignment was active and only to projects associated with that event.

Manual sharing proved unreliable. Access was often left in place long after assignments ended, creating compliance risks.

Apex-managed sharing allowed the system to

- Grant access when a contractor's assignment started

- Revoke access automatically when the assignment end date passed

- Ensure no human intervention was required

Time-bound logic like this cannot be enforced consistently through sharing rules. Programmatic control becomes essential.

Scenario 3: Conditional Access Based on Risk Classification

Risk-based access models are increasingly common in regulated and mission-critical environments. Records are not simply public or private; they exist along a spectrum of sensitivity.

Imagine a global humanitarian organization tracking aid delivery projects. Projects are classified as follows:

- Low risk

- Medium risk

- High risk

Risk classification depends on multiple inputs:

- Country stability index

- Funding source type

- Past audit outcomes

When a project crosses into "High Risk" status, additional oversight is required. Senior compliance officers must gain immediate visibility, even if they are outside the project's normal reporting line.

Sharing rules struggle here because

- Risk classification is derived, not static.

- Oversight roles may change over time.

- Access must be revoked if risk decreases.

Apex-managed sharing allows access to follow the *state* of the record, not just its ownership or category.

Scenario 4: Matrix Organizations with Skill-Based Visibility

Matrix organizations challenge traditional hierarchy-based access. Visibility is often determined by skills, certifications, or temporary team membership rather than reporting lines.

Consider a global engineering firm using Salesforce to manage complex client engagements. Specialists are assigned to projects based on expertise:

- Cybersecurity

- Cloud architecture

- Regulatory compliance

Assignments change frequently, and specialists may work across multiple regions simultaneously.

Declarative sharing assumes relatively stable ownership or role-based access. In contrast, Apex-managed sharing can respond to

- Assignment records being created or removed

- Skill certifications expiring

- Temporary task forces being formed and dissolved

Visibility follows *capability*, not position.

Scenario 5: Access Controlled by External Systems

In some architectures, Salesforce is not the system of record for access decisions. External systems—such as ERP platforms, grant management tools, or identity providers—determine who should see what.

A global foundation integrated Salesforce with an external grant lifecycle system. Grant approvals, suspensions, and terminations occurred outside Salesforce, but record visibility needed to reflect those decisions immediately.

Sharing rules could not respond directly to external events. Manual sharing was impractical at scale.

Apex-managed sharing acted as the translation layer:

- External events triggered access updates.

- Share records were created or removed based on authoritative decisions.

- Salesforce remained aligned without duplicating logic.

This pattern is increasingly common in enterprise architectures.

Scenario 6: Gradual Access Expansion During Process Lifecycles

Many business processes unfold in stages, with visibility expanding as commitments increase.

For example, in a public-sector procurement process

- Early-stage proposals are visible only to evaluators.

- Shortlisted proposals become visible to legal reviewers.

- Awarded proposals are visible to finance and delivery teams.

The same record moves through multiple access states as it progresses.

Declarative sharing rules struggle to model this progression cleanly, especially when transitions must trigger both access grants and revocations.

Apex-managed sharing allows visibility to evolve alongside the process itself, ensuring that access remains aligned with responsibility.

Scenario 7: Legal Holds and Investigative Access

Occasionally, access requirements arise that are explicitly exceptional and sensitive.

During internal investigations or legal holds,

- Specific records must be made visible to a small group.

- Visibility must be tightly controlled and auditable.

- Access must be removed immediately when the matter concludes.

These situations demand precision and traceability. Sharing rules are too broad; manual sharing is too fragile.

Apex-managed sharing allows access to be granted under a clearly documented, auditable reason and revoked with certainty.

Scenario 8: Multi-Tenant Community Models

In Experience Cloud implementations, data is often shared with external users under strict conditions.

Consider a partner portal where multiple organizations collaborate on shared initiatives, but

- Each partner must see only their own records.

- Joint initiatives require selective cross-visibility.

- Access rules depend on contractual relationships stored in Salesforce.

Declarative sharing quickly becomes unmanageable in these environments. Apex-managed sharing enables tenant-aware visibility that respects contractual boundaries.

Lessons from the Field

Across these scenarios, several themes emerge:

- **Context matters more than structure.**

- **Relationships matter more than ownership.**

- **Time matters as much as hierarchy.**

Apex-managed sharing excels when access decisions are

- Conditional

- Temporal

- Derived

- Contextual

It is not a tool for convenience, but for correctness.

Designing for Reality, Not Simplicity

Real organizations are messy. People wear multiple hats. Processes evolve. Exceptions become rules.

Apex-managed sharing acknowledges this reality. It allows architects to model access as it actually exists, rather than forcing business behavior to conform to system limitations.

The key is intentional use. Apex-managed sharing should appear where it adds clarity—not where it merely adds power.

Bridging to Performance Considerations

As these real-world scenarios demonstrate, Apex-managed sharing often operates at scale, under complexity, and with high expectations for correctness.

This makes performance considerations unavoidable.

In the next section, we will explore **performance implications and best practices**, focusing on how to ensure Apex-managed sharing remains efficient, predictable, and safe—even in large, high-volume Salesforce environments.

Performance Considerations—Making Apex-Managed Sharing Scale Safely

Apex-managed sharing offers precision, but that precision comes at a cost. Every custom sharing decision ultimately affects Salesforce's core visibility engine. When misused or poorly designed, Apex-managed sharing can quietly become one of the most expensive operations in an org—both in system resources and operational risk.

Performance, in this context, is not just about speed. It is about **predictability**, **resilience**, and **long-term sustainability** as data volumes and business complexity grow.

Understanding the Cost of Sharing Calculations

Every time Salesforce evaluates record access, it consults multiple layers:

- Organization-Wide Defaults

- Role hierarchy

- Sharing rules

- Manual shares

- Apex-managed share records

Apex-managed shares are evaluated alongside other sharing mechanisms, but unlike declarative sharing, they must be explicitly created, maintained, and removed through code. This means poorly designed custom sharing logic can degrade performance far beyond the moment when the share record is created.

166

The performance impact shows up in three primary ways:

- Slower queries

- Longer transaction times

- Increased background recalculation workload

In small orgs, this impact may go unnoticed. In large, data-heavy environments, it becomes impossible to ignore.

Volume Is the First Performance Multiplier

The most common mistake teams make is testing Apex-managed sharing in low-volume environments and assuming the results will scale.

A pattern that works perfectly for

- 500 records

- 50 users

- A handful of groups

may fail catastrophically at

- Five million records

- Thousands of users

- Dynamic public group membership

Each additional share row increases the complexity of access evaluation. Unlike many other Salesforce features, sharing does not benefit significantly from caching. Visibility checks are performed continuously and must remain accurate at all times.

The first performance question architects should ask is not *"Does this logic work?"* but *"How many share rows will this create over time?"*

Share Explosion: The Silent Performance Killer

"Share explosion" occurs when a single business rule generates a disproportionately large number of share records.

Common causes include the following:

- Sharing one record with many users

- Sharing many records with one user

- Sharing many records with many users simultaneously

In isolation, none of these patterns is inherently wrong. The danger arises when they compound.

For example, sharing every active project with every certified specialist might seem reasonable—until certifications change weekly and projects number in the hundreds of thousands. The system spends more time maintaining visibility than supporting business operations.

Apex-managed sharing demands careful estimation of worst-case scenarios, not just typical usage.

Why Recalculation Matters More Than Insertion

Most teams focus performance testing on the moment share records are inserted or deleted. While this matters, it is often not where the real cost lies.

The heavier cost is **ongoing recalculation**:

- When users run reports.

- When dashboards refresh.

- When integrations query data.

- When batch jobs execute.

Each of these operations triggers visibility checks. The more complex the sharing model, the more expensive those checks become.

This is why performance issues related to sharing often surface far from the original logic—in reporting delays, API timeouts, or sluggish user interfaces.

Timing Matters: Synchronous vs. Asynchronous Sharing

One of the most important architectural decisions is *when* sharing logic executes.

Running Apex-managed sharing synchronously—during record creation or update—can increase transaction time and risk hitting governor limits. In high-volume environments, this can result in failed transactions or unpredictable user experiences.

Many mature implementations separate:

- **Business logic** (what changed)

- **Sharing logic** (who should now see it)

By deferring sharing updates to asynchronous processes, organizations reduce user-facing latency while maintaining accuracy.

Performance is not just about speed; it is about isolating heavy work from critical paths.

Bulk Behavior Is Non-negotiable

Salesforce is inherently bulk-oriented. Any Apex-managed sharing logic that assumes single-record execution will eventually fail.

Performance degradation often begins subtly:

- A nightly job runs slightly longer.

- A data load takes an extra hour.

- A deployment introduces intermittent failures.

These are early warning signs of non-bulk-safe design.

Even when sharing logic appears simple, it must assume:

- Hundreds of records entering or exiting eligibility at once

- Multiple users gaining or losing access simultaneously

- Reprocessing during data corrections or backfills

Bulk safety is not an optimization—it is a requirement.

The Hidden Cost of Revocation

Granting access is only half the story. Revoking access often costs more.

Deletion of share records

- Requires identifying the correct rows

- Triggers recalculation

- Can cascade if multiple rules overlap

Many teams discover performance issues only after introducing revocation logic, especially when criteria fluctuate frequently.

For example, access tied to a "status" field that changes often can result in constant churn:

- Grant

- Revoke

- Grant again

This oscillation increases system load without adding business value. Stable criteria and thoughtful thresholds reduce unnecessary recalculations.

Designing for Stability Over Reactivity

High-performing sharing models favor stability. This means

- Avoiding criteria that change frequently

- Preferring milestone-based transitions over real-time sensitivity

- Accepting small delays in access changes when appropriate

Not every access decision needs to be immediate. In many cases, slight delays—minutes or even hours—are acceptable and dramatically improve system health.

Performance-conscious architects design sharing logic that reacts *intentionally*, not reflexively.

Reporting and Analytics: The Canary in the Coal Mine

Reporting performance often reveals sharing problems before anything else.

Symptoms include the following:

- Dashboards timing out

- Reports running inconsistently

- API-based analytics slowing down

Because reports evaluate visibility across large datasets, they magnify the cost of complex sharing models. Monitoring reporting performance is one of the most effective ways to detect sharing-related issues early.

Organizations that treat reporting delays as purely "analytics problems" often miss deeper architectural causes.

Monitoring and Observability

Unlike many Salesforce features, sharing behavior is difficult to observe directly. There is no native dashboard that shows the cost of your sharing model.

High-performing organizations compensate by

- Tracking the growth of share tables over time

- Monitoring batch job durations

- Watching for increases in query execution time

- Correlating performance changes with sharing logic updates

Performance governance requires visibility—even if that visibility must be inferred indirectly.

Performance Trade-Offs Are Architectural Decisions

There is no free lunch in Apex-managed sharing. Every decision introduces trade-offs:

- Precision vs. simplicity

- Real-time accuracy vs. system load

- Granularity vs. scalability

The role of the architect is not to eliminate trade-offs, but to choose them consciously.

In many cases, slightly broader access with fewer rules performs better—and is safer—than hyper-precise sharing that overwhelms the platform.

Lessons Learned from Large-Scale Implementations

Organizations that successfully use Apex-managed sharing at scale tend to follow consistent patterns:

- They limit the number of custom sharing dimensions.

- They centralize sharing logic conceptually, even if implemented across processes.

- They test performance with realistic data volumes, not samples.

- They review sharing models as part of regular architecture governance.

Most importantly, they treat sharing as infrastructure—not as a feature added reactively to individual requests.

The Governance Layer

Performance and governance are inseparable. A fast but poorly governed sharing model will eventually fail. Likewise, a well-governed model that ignores performance will collapse under growth.

In the next section, we will explore **governance strategies for Apex-managed sharing**—focusing on ownership, auditability, compliance, and long-term maintainability.

Governance: Sustaining Trust, Compliance, and Control in Apex-Managed Sharing

As Salesforce implementations mature, governance becomes the quiet force that determines whether a sharing model continues to serve the organization—or slowly turns into a liability. This is especially true for Apex-managed sharing. Unlike declarative sharing rules, which are visible and centrally configured, programmatic sharing operates behind the scenes. Without strong governance, even well-intentioned logic can lead to invisible access paths, audit challenges, and long-term maintenance risks.

Governance in the context of Apex-managed sharing is not about restricting developers. It is about ensuring that **every access decision is explainable, reviewable, and reversible**—not just today, but years into the future.

This section explores how to put guardrails around custom sharing logic so that it remains aligned with business intent, regulatory expectations, and organizational change.

Why Governance Matters More for Apex-Managed Sharing

Declarative sharing tools naturally encourage governance. Sharing rules are named, documented in Setup, and reviewed during audits or admin handovers. Apex-managed sharing, however, bypasses much of that visibility. Access can be granted dynamically, at scale, and based on logic that may not be obvious to administrators or compliance teams.

The risks are subtle but real:

- Users may have access to records without anyone remembering why.

- Legacy logic may continue running long after the business process has changed.

- Auditors may struggle to understand how and why sensitive data is visible.

- New team members may hesitate to touch sharing logic they do not fully understand.

Governance does not eliminate these risks entirely, but it reduces them to a manageable level by making intent explicit and behavior predictable.

Establishing Clear Ownership of Sharing Logic

One of the most common governance failures is unclear ownership. Apex-managed sharing often spans multiple teams—developers implement it, admins support it, security teams audit it, and business stakeholders depend on it. When responsibility is diffused, accountability disappears.

A sustainable model assigns **clear ownership** at multiple levels:

- **Technical Owner**: Responsible for the code structure, performance, and maintainability

- **Business Owner**: Responsible for defining who should see what and why

- **Governance Owner**: Responsible for periodic reviews and compliance alignment

These roles may be held by different individuals or the same person, but they must be explicitly identified. When a question arises—*Why does this user have access?*—there should be no ambiguity about who can answer it.

Making Access Decisions Explainable

A guiding principle of governance is explainability. Every share row created through Apex should be traceable back to a business reason that a nontechnical stakeholder can understand.

This is where **intent documentation** becomes as important as technical design. For every Apex-managed sharing use case, governance documentation should answer

- What business problem does this access solve?

- Who requested it, and under what conditions?

- Is the access permanent, conditional, or time-bound?

- What event should remove the access?

This documentation does not need to be elaborate, but it must exist outside the code itself. Relying on comments or developer memory is not governance—it is technical debt waiting to surface.

Row Cause Governance As an Audit Anchor

RowCause values play a central role in governance because they act as the **label on every access decision**. When used correctly, they allow administrators and auditors to distinguish between different sources of access without reverse-engineering logic.

From a governance perspective, RowCause values should be treated as **controlled vocabulary**, not casual identifiers. Each custom RowCause should have the following:

- A clearly defined business meaning

- A documented lifecycle (when it is created and removed)

- A single owning use case

Reusing a RowCause across unrelated scenarios may seem efficient in the short term, but it undermines audit clarity. When auditors see thousands of share rows with the same reason, they should be able to understand exactly what that reason represents—without exceptions or footnotes.

Aligning Sharing Governance with Compliance Expectations

In regulated environments, governance is not optional. Whether driven by internal policy or external regulation, organizations must demonstrate control over who can access sensitive data and why.

Apex-managed sharing intersects with compliance in three critical ways:

1. **Visibility Justification:**

   It must be possible to explain why a user can see a particular record at a particular time.

2. **Change Traceability:**

   When access rules change, there should be a record of what changed and why.

3. **Periodic Validation:**

   Access granted through custom logic must be reviewed just as rigorously as profile or permission set assignments.

Governance frameworks often focus heavily on configuration but overlook code-driven access. Mature organizations explicitly include Apex-managed sharing in their access reviews and compliance checklists.

Periodic Review As a Governance Practice

One of the most effective governance tools is also the simplest: **regular review**. Apex-managed sharing logic should not be treated as "set and forget." Business processes evolve, organizational structures change, and assumptions that were valid two years ago may no longer hold.

A practical governance cadence includes the following:

- Quarterly or biannual review of active sharing scenarios

- Validation that business conditions triggering access still exist

- Confirmation that revocation logic still executes as expected

- Cleanup of deprecated logic and unused RowCause values

These reviews are not about rewriting code—they are about reaffirming intent. Often, the outcome is reassurance rather than change, but even that reassurance has value.

Managing Change Without Breaking Trust

Change is inevitable in Salesforce implementations. New objects are introduced, fields are renamed, processes are re-engineered, and teams are reorganized. Apex-managed sharing must be resilient to this change, or governance will erode over time.

Strong governance encourages **change impact awareness**. Before modifying fields, relationships, or automation that influence sharing logic, teams should ask:

- Does this affect how access is granted or revoked?

- Will this introduce unintended visibility?

- Should existing share rows be recalculated or cleaned up?

This mindset prevents accidental exposure and reinforces trust between technical teams and business stakeholders.

Governance Through Transparency, Not Obscurity

A common misconception is that security improves when systems become harder to understand. In practice, the opposite is true. The most secure sharing models are those that are **transparent, well-documented, and openly reviewed**.

Apex-managed sharing should never feel like a black box. Admins should be able to answer basic questions about access without diving into code. Security teams should be able to audit sharing behavior without guesswork. Business leaders should feel confident that visibility aligns with intent.

Governance is what makes this possible.

A Governance Mindset for Architects

For Salesforce architects, governance is not an afterthought—it is a design principle. The question is not only *Can we implement this sharing logic?* but also

- Will someone understand this five years from now?

- Can we confidently explain this to an auditor?

- Can this logic evolve without creating hidden risks?

When Apex-managed sharing is governed well, it becomes a strategic asset rather than a fragile customization.

Key Takeaways

- Apex-managed sharing demands stronger governance than declarative models.

- Clear ownership and documentation are non-negotiable.

- RowCause values are central to auditability and clarity.

- Regular review prevents silent drift and access creep.

- Transparency builds long-term trust in the sharing model.

Governance does not slow teams down—it allows them to move forward without losing control. In advanced sharing architectures, it is the difference between flexibility and fragility.

BrainstormingDesigning Apex-Managed Sharing with Intent

Apex-managed sharing is rarely about technology alone. It sits at the intersection of trust, visibility, accountability, and organizational reality. This section is designed to slow down—to step away from configuration screens and code editors—and think like an architect.

The goal of these exercises is not to produce a single "correct" design. Instead, they encourage deliberate reasoning about **why access exists, who it serves**, and **how it should evolve over time**.

Use these prompts individually, in workshops, or as part of design reviews. They are especially valuable when decisions feel complex or politically sensitive—because those are the moments when poor sharing decisions tend to linger longest.

Reframing the Core Question: "Who Should See This?"

Most sharing conversations begin with a narrow request:

"This team needs access to those records."

A more useful architectural question is broader:

"Under what conditions should this information be visible—and when should it not be?"

Reflection Prompt

Think about a custom object in your org that required advanced sharing logic.

- What event caused the sharing requirement to surface?

- Was the request driven by a permanent need or a temporary workaround?

- If the same request were made today, would you solve it the same way?

Often, Apex-managed sharing emerges not because the data model is complex, but because the **decision-making context** around access is complex.

Identifying the True Drivers of Visibility

A common mistake is assuming that record ownership or static field values are the primary drivers of access. In advanced implementations, visibility is often influenced by **relationships**, **timing**, and **risk posture**.

Exercise: Visibility Drivers Map

For a given object, list all possible factors that might influence access

- Record attributes (status, classification, sensitivity)

- Related records (parent objects, agreements, assignments)

- User attributes (role, skill set, certification)

- Time-based factors (project phase, contract duration)

- External considerations (compliance, donor restrictions, audits)

Now ask

- Which of these can be handled declaratively?

- Which require context-aware evaluation?

- Which should *not* drive access, even if technically possible?

This exercise often reveals that Apex-managed sharing is justified not by one factor, but by the **interaction between several factors**.

Separating Business Intent from Technical Mechanism

One of the most important architectural skills is the ability to describe access rules without referencing Salesforce features.

Thought Experiment

Describe a sharing requirement **without** using terms like

- Sharing rule

- Role hierarchy

- Apex

- Permission set

Instead, frame it in business language

- Who needs access?

- To what?

- Under which conditions?

- For how long?

- With what level of responsibility?

If the requirement cannot be clearly explained without technical terms, it is often a sign that the logic is either under-defined or over-engineered.

Deciding When Apex Is Truly the Right Tool

Apex-managed sharing should feel *inevitable*, not convenient.

Discussion Starters

For a proposed sharing solution, ask

- Can this requirement be satisfied if we slightly adjust the data model?

- Are we compensating for unclear ownership or process gaps?

- Will this logic still make sense if the organization doubles in size?

- What happens if this logic fails silently?

If Apex is chosen simply because it feels more flexible, governance problems usually follow.

Designing for Removal, Not Just Access

Many teams focus intensely on how access is granted—but give little thought to how it is removed. This is one of the most common sources of overexposure.

Exercise: Access Lifecycle Walkthrough

Choose one Apex-managed sharing scenario and walk through its full lifecycle

- What event grants access?

- What conditions keep access active?

- What event should remove access?

- What happens if that removal event never occurs?

- Who notices?

This exercise often reveals gaps where access persists longer than intended—not because of malicious intent, but because revocation was never fully defined.

Balancing Flexibility with Predictability

Apex-managed sharing is powerful precisely because it is flexible. But flexibility without predictability erodes trust.

Reflection Prompt

Consider two users with identical roles and profiles.

- Can you confidently explain why one sees a record and the other does not?

- Can an administrator determine that reason without reading code?

- Would an auditor accept that explanation?

If the answers are uncertain, the design may need stronger governance—not stronger logic.

Evaluating Risk Through a Sharing Lens

Advanced sharing decisions often involve sensitive or high-impact data.

Scenario Exercise

Imagine a record type that contains

- Financial commitments

- Personal data

- Legal assessments

- Strategic forecasts

Now consider

- What is the cost of overexposure?

- What is the cost of underexposure?

- Which risk is more damaging in your context?

Apex-managed sharing is often chosen when underexposure blocks critical work—but governance must ensure that solving one risk does not amplify another.

Designing for Organizational Change

Organizations rarely remain static. Teams merge, regions reorganize, and priorities shift.

Thought Exercise

Assume that within two years

- Team structures will change.

- Job titles will be renamed.

- Some processes will be retired.

Ask yourself

- Will this sharing logic adapt gracefully?

- Or will it hard-code today's org chart into tomorrow's system?

Architectural maturity is measured not by how well a system handles today's requirements, but by how calmly it absorbs tomorrow's change.

Group Workshop Activity: Sharing Design Review

This activity works well in team settings.

3. Select one Apex-managed sharing use case.

4. Assign roles:

   - Business stakeholder

   - Salesforce admin

   - Security/compliance reviewer

   - Architect

5. Have each role explain the sharing logic from their perspective. Common outcomes:

   - Business stakeholders focus on outcomes, not mechanics.

   - Admins focus on visibility and troubleshooting.

- Security reviewers focus on justification and auditability.

- Architects focus on alignment and longevity.

Where these perspectives clash is where design improvements usually emerge.

Final Reflection: The Architect's Responsibility

Apex-managed sharing is not just a technical feature—it is a statement of trust. It says

- We trust the system to make nuanced decisions.

- We trust future teams to understand those decisions.

- We trust governance to keep flexibility from becoming chaos.

The best sharing architectures are rarely the most complex. They are the ones where **every access path exists for a reason—and that reason can be clearly articulated**.

Closing Thoughts

If declarative sharing defines *what is allowed*, Apex-managed sharing defines *what is possible*.

Your responsibility as an architect is to ensure that what is possible remains appropriate, explainable, and controlled—long after the original requirement has faded from memory.

CHAPTER 6

Optimizing and Troubleshooting Your Sharing Model

Optimizing and troubleshooting sharing is not about adding more rules or writing more logic. It is about understanding how visibility is actually being evaluated, recognizing patterns that no longer scale, and making deliberate architectural decisions to keep access predictable, performant, and auditable. Many production issues attributed to "Salesforce being slow" or "users not seeing records" are in reality symptoms of a sharing model that has outgrown its original design.

Geared toward administrators, architects, developers, and governance/compliance leads, this chapter is structured to mirror how experienced architects and administrators approach sharing challenges in the real world. Understanding what Salesforce is actually doing-rather than what we assume it is doing-is the foundation for every improvement. This chapter is intentionally practical yet conceptual and focuses on what happens *after* a sharing model has been implemented. You will not find step-by-step configuration instructions or code listings. Instead, the focus is on *how to think* about sharing at scale.

Laying the Groundwork

As Salesforce implementations mature, sharing models often become one of the most complex—and least visible—parts of the platform. Early in an organization's Salesforce journey, record visibility tends to be straightforward. Teams rely on role hierarchies, Organization-Wide Defaults (OWDs), and a handful of sharing rules. Over time, however, business processes evolve. New teams are introduced, compliance expectations tighten, and data volumes increase. What once worked cleanly can begin to feel fragile, slow, or opaque.

© Sandhya Sharma 2026

S. Sharma, *Unlocking Salesforce Data*, https://doi.org/10.1007/979-8-8688-2305-3_6

If you are:

- An **administrator**, this chapter will help you diagnose access issues with greater confidence.

- An **architect**, it will support you in making defensible design decisions under real-world constraints.

- A **developer**, it will sharpen your understanding of when programmatic sharing helps—and when it hurts.

- A **governance or compliance lead**, it will clarify how visibility can be monitored and explained.

A well-designed sharing model should meet three core goals simultaneously:

- **Correctness**: Users see exactly the records they should—no more and no less.

- **Performance**: Visibility checks do not degrade query speed, report execution, or background processing.

- **Governance**: Access can be explained, audited, and adjusted without destabilizing the system.

When any one of these pillars is neglected, the effects ripple outward. Support teams spend time diagnosing access issues instead of solving business problems. Administrators hesitate to make changes because the impact is unclear. Architects are forced into reactive fixes rather than strategic improvements. This chapter equips you to move from reactive troubleshooting to proactive optimization. Most importantly, it will help you move from reactive fixes to intentional design.

Why Sharing Models Degrade Over Time

Sharing models rarely fail overnight. Instead, they accumulate complexity gradually, often in response to legitimate business needs.

A new region requires temporary access to records. A compliance audit introduces additional visibility restrictions. A merger brings new roles and reporting lines. Each change, taken in isolation, may be reasonable. Over years, however, the combined effect can be difficult to reason about.

Common indicators that a sharing model needs attention include the following:

- Users intermittently losing access after record updates.

- Reports taking noticeably longer to run as data volumes grow.

- Administrators unsure which rule or mechanism is granting access.

- Apex-managed shares that are never revoked.

- Visibility logic duplicated across multiple triggers or processes.

These are not failures of Salesforce as a platform. They are signs that the sharing architecture was not revisited as the organization scaled.

Optimization, therefore, is not about "fixing mistakes" but about *realigning the sharing model with current realities.*

The Hidden Cost of Poor Visibility Design

Unlike validation rules or automation failures, sharing issues are often silent. Salesforce does not raise an error when a user cannot see a record they should. From the system's perspective, everything is working as designed.

The cost is paid elsewhere:

- **Operational Inefficiency**: Support tickets multiply as users request access that "used to work."

- **Compliance Risk**: Excessive access is harder to detect than missing access, especially when multiple sharing mechanisms overlap.

- **Performance Drag**: Each additional sharing entry increases the work Salesforce must do to resolve visibility during queries.

- **Architectural Inertia**: Teams avoid refactoring sharing because the rules feel too risky to touch.

Over time, organizations may accept these issues as unavoidable. In reality, many can be mitigated—or eliminated—through disciplined monitoring, auditing, and architectural refinement.

Optimization As an Ongoing Practice

One of the most important mindset shifts for advanced Salesforce practitioners is recognizing that sharing is not a "set once" configuration. It is a living system that must be observed and adjusted.

Optimization involves asking questions such as

- Which sharing mechanisms are actively used today?

- Which exist only because they were added years ago?

- Are we relying on Apex-managed sharing where declarative rules would now suffice?

- Do we know why each share exists—and when it should be removed?

Troubleshooting, in this context, is not a sign of failure. It is feedback. Visibility issues reveal mismatches between business expectations and technical reality. Treated thoughtfully, they become opportunities to simplify and strengthen the model.

Monitoring and Auditing Sharing Settings

Most Salesforce teams invest significant time designing their sharing model. Organization-Wide Defaults are debated, role hierarchies are drawn on whiteboards, and edge cases are carefully considered. Once the model is implemented and appears to work, attention usually shifts elsewhere. Visibility becomes a "set it and forget it" concern.

That assumption is one of the most common causes of long-term sharing problems.

Why Monitoring Matters More Than Configuration

Sharing is not static. Users change roles, business processes evolve, compliance requirements tighten, and data volumes grow. A sharing model that was perfectly acceptable at launch can quietly drift out of alignment with reality. Over time, access may become too broad, too restrictive, or simply opaque—where no one is entirely sure why a user can or cannot see a particular record.

Monitoring and auditing are therefore not optional administrative tasks. They are an essential part of maintaining trust in the system. When users trust that Salesforce shows them exactly what they are entitled to see—and nothing more—they rely on it more heavily, make better decisions, and escalate fewer issues.

At an organizational level, effective monitoring also protects against risk. Data visibility is inseparable from compliance. Whether the concern is donor confidentiality, customer privacy, internal controls, or contractual boundaries, the ability to explain *who can see what and why* is no longer a nice-to-have.

Understanding What "Monitoring" Really Means in Sharing

Monitoring a sharing model is not the same as monitoring system performance or user logins. It is a more subtle discipline.

At its core, sharing monitoring answers three ongoing questions as follows:

- **Who currently has access to a record?**

- **Through which mechanism was that access granted?**

- **Does that access still make sense today?**

The challenge is that Salesforce can grant access through multiple overlapping paths. A single user might see a record because they own it, sit above the owner in the role hierarchy, belong to a public group, qualify for a sharing rule, and receive temporary access through Apex-managed sharing. From the user's perspective, access simply "exists." From an administrator's perspective, it must be explainable.

Effective monitoring does not try to eliminate complexity. Instead, it creates visibility into complexity.

The Layers of Sharing That Require Oversight

A practical way to approach auditing is to think in layers rather than mechanisms.

The baseline layer is defined by Organization-Wide Defaults and role hierarchy. This layer establishes the minimum and maximum exposure for an object. Monitoring here focuses on whether the baseline still aligns with business intent. For example, an object that was once intentionally public may no longer need to be, or a private model may be forcing excessive custom sharing downstream.

The declarative layer includes sharing rules and manual sharing. These tend to grow organically over time. New rules are added to address new use cases, but old ones are rarely removed. Auditing this layer involves reviewing not just what rules exist, but whether they are still actively serving a purpose.

The programmatic layer—Apex-managed sharing—is where monitoring becomes most critical. Unlike declarative rules, programmatic sharing is often invisible to administrators unless they know exactly where to look. Share records can persist long after the business condition that created them has disappeared.

The exception layer includes special cases such as delegated administration, territory management, and implicit access through related records. These are frequently overlooked during audits, yet they can significantly affect visibility.

A healthy monitoring strategy touches all four layers, even if the emphasis varies depending on the organization's complexity.

Making Sharing Explainable to Humans

One of the clearest signs of a mature sharing model is how easily it can be explained to a nontechnical stakeholder.

If a data protection officer, auditor, or business owner asks why a specific user can see a sensitive record, the answer should not require guesswork or trial and error. Monitoring exists to ensure that answers are available, traceable, and defensible.

This is where documentation and metadata discipline become as important as technical correctness. Clear naming of public groups, descriptive sharing rule names, and well-defined Row Causes in Apex-managed sharing all contribute to explainability.

Many experienced architects treat explainability as a first-class requirement. If a sharing mechanism cannot be easily explained, it is often redesigned—not because it fails technically, but because it fails operationally.

Auditing Share Objects As System Records

From Salesforce's perspective, share objects are system data. From an auditor's perspective, they are evidence.

Every row in a share object represents a deliberate decision to grant access. Over time, these rows form a historical footprint of how visibility has evolved. Auditing involves periodically reviewing this footprint to identify patterns that no longer align with policy.

Common audit questions include the following:

- Are there large volumes of share records associated with a single user or group?

- Do certain Row Causes dominate the share table in unexpected ways?

- Are there share records tied to inactive users, expired teams, or obsolete processes?

Audits are not about finding faults; they are about identifying drift. Most issues uncovered during sharing audits are the result of good intentions combined with changing realities.

Temporal Access and the Problem of "Forgotten Visibility"

One of the most common findings during audits is lingering access.

Temporary access is easy to grant and easy to forget. Contractors join a project, emergency teams are assembled, cross-functional visibility is opened for a short-term initiative—and then the initiative ends. Unless revocation is built into the process, access quietly remains.

Monitoring strategies should therefore include time as a dimension. Access that made sense six months ago may be actively harmful today. Mature organizations treat time-bound access as a lifecycle, not an event.

Some teams formalize this through periodic reviews. Others align sharing audits with business milestones, such as fiscal year-end or program closure. The method matters less than the discipline of revisiting assumptions.

Visualizing Sharing for Better Insight

Many teams find that visual representations of sharing paths dramatically improve understanding. Diagrams that show how access flows—from ownership to hierarchy, to rules, to custom logic—help surface unintended overlaps.

These visuals are especially valuable during audits, workshops, and compliance reviews. They turn abstract configuration into something concrete and discussable.

Even a simple diagram that distinguishes "baseline access" from "exceptions" can reveal where complexity has accumulated unnecessarily.

Monitoring As an Ongoing Practice, Not an Event

Perhaps the most important mindset shift is to treat sharing audits as routine rather than reactive.

Visibility issues often surface only after a problem occurs: a user sees data they should not, or cannot see data they urgently need. By the time the issue is reported, the root cause may be buried under layers of historical configuration.

Regular monitoring reduces surprises. It transforms sharing from a fragile construct into a resilient system that adapts as the organization grows.

Teams that succeed in this area do not rely on heroics or institutional memory. They rely on repeatable practices, clear ownership, and a shared understanding that visibility is as critical as data quality itself.

Lessons Learned from the Field

Monitoring and auditing sharing settings is not about control for its own sake. It is about confidence—confidence that Salesforce reflects organizational intent, respects boundaries, and can stand up to scrutiny.

When monitoring is done well, troubleshooting becomes faster, scaling becomes safer, and compliance becomes less stressful. It lays the foundation for every other optimization discussed in this chapter.

In the next section, we move from prevention to response, examining how to systematically troubleshoot visibility issues when expectations and reality diverge.

Troubleshooting Common Visibility Issues

Even the most thoughtfully designed sharing model will eventually face questions, complaints, or outright confusion from users. "I can't see this record anymore." "Why does my colleague have access but I don't?" "This was visible yesterday—what changed?" These questions are not signs of a broken system; they are signs of a living one.

Troubleshooting sharing issues in Salesforce is less about quick fixes and more about disciplined analysis. The platform evaluates record visibility through a layered, cumulative model, and understanding how those layers interact is essential to diagnosing problems accurately. This section focuses on how to think through visibility issues methodically, identify root causes, and resolve them without destabilizing the broader security architecture.

Understanding Visibility As a Layered Outcome

One of the most common mistakes administrators and developers make is treating record visibility as a single decision point. In reality, visibility is the result of multiple independent mechanisms working together.

At a high level, Salesforce evaluates access conceptually in this order:

- Object-level permissions

- Organization-Wide Defaults

- Record ownership and role hierarchy

- Declarative sharing rules

- Team-based access (where applicable)

- Manual sharing

- Apex-managed sharing

A failure at any earlier layer cannot be compensated for by a later one. For example, no amount of Apex-managed sharing will help if the user lacks read access to the object itself. Effective troubleshooting always begins by confirming that the foundational layers are intact before examining more advanced logic.

Issue 1: "I Should Have Access, But I Don't"

This is the most frequently reported issue and often the most misleading. Users typically assume access based on business context ("I work on this project") rather than system logic.

The first diagnostic step is always object-level access. Profiles and permission sets define whether a user can read or edit a given object at all. If access has recently changed, the cause may be as simple as a permission set removal or profile reassignment.

Once object access is confirmed, ownership and hierarchy should be examined. In private sharing models, users often assume that being "senior" or "involved" implies visibility, when in fact access flows only through explicit hierarchy or sharing rules. A user who has changed roles or moved teams may unknowingly lose inherited access.

If the issue persists, attention should shift to sharing rules and custom logic. Criteria-based rules can fail silently when a field value changes or when record ownership moves to a queue or integration user. Apex-managed sharing adds further complexity, as access may depend on conditions evaluated at a specific moment in time rather than continuously.

Issue 2: "Someone Else Can See It, But I Can't"

This scenario is particularly useful diagnostically because it provides a comparison point. When two users appear similar from a business perspective but have different access, the difference is almost always structural.

Key areas to compare include the following:

- Role assignments and hierarchy position

- Permission sets and permission set groups

- Team memberships

- Public group inclusion

- Temporary access mechanisms, such as manual shares or time-based logic

In organizations using Apex-managed sharing, differences often stem from contextual conditions. One user may meet eligibility criteria based on geography, certification, or project assignment, while another does not—even if both believe they "should" have access. These situations underscore the importance of documenting sharing logic in business terms, not just technical ones.

Issue 3: "I Had Access Before—Now It's Gone"

Access revocation is frequently more confusing than access grants. Users tend to notice loss more acutely than gain, and the system rarely provides an obvious explanation.

Common causes include the following:

- Field changes that invalidate criteria-based sharing

- Ownership transfers

- Automated cleanup of Apex-managed shares

- Role or group membership changes

- End of time-bound access (such as contractor engagements)

From a troubleshooting perspective, this symptom highlights the importance of reversibility awareness. Well-designed sharing models intentionally remove access when conditions no longer apply. The challenge lies in distinguishing intentional revocation from accidental loss.

Administrators should resist the temptation to "just add a manual share" to resolve the complaint. Doing so may mask a deeper issue or undermine carefully designed governance controls.

The Role of Row Cause in Diagnostics

For organizations using Apex-managed sharing, the RowCause field becomes one of the most valuable diagnostic tools available. Each share record includes a reason explaining why access exists. When custom RowCauses are clearly named and consistently used, they tell a story.

For example:

- A RowCause indicating regulatory oversight explains access that would otherwise seem excessive.

- A RowCause tied to a project phase explains why access disappeared after a stage change.

- A RowCause linked to temporary collaboration clarifies why access is time-limited.

When troubleshooting, reviewing share records and their RowCauses often reveals whether access was granted intentionally, inherited indirectly, or applied temporarily. Without this clarity, teams are left guessing—and guessing leads to risky fixes.

Visibility Issues Caused by Data Changes

A subtle but frequent source of visibility problems is data evolution. Sharing rules and Apex logic are often written against fields assumed to be stable. Over time, those assumptions break.

Examples include the following:

- New picklist values introduced without updating sharing logic

- Lookup relationships repointed during data cleanup

- Integration processes overwriting fields used in sharing criteria

- Record types added without corresponding sharing consideration

Troubleshooting in these cases requires collaboration between administrators, developers, and business stakeholders. The issue is not "broken sharing" but outdated assumptions encoded into the model.

Diagnosing Issues in Large or Mature Orgs

In mature Salesforce environments, visibility issues are rarely isolated. Multiple sharing mechanisms may apply simultaneously, sometimes redundantly and sometimes in conflict.

Effective troubleshooting in such environments depends on

- Clear documentation of sharing intent

- Consistent naming conventions for groups and rules

- A shared understanding of which mechanism "owns" which use case

- Resisting ad hoc fixes that bypass established patterns

Experienced teams often maintain a mental map of their sharing architecture. Less-experienced teams rely on trial and error. The difference becomes apparent when issues arise at scale.

When the System Is Working As Designed—but Expectations Are Not

Perhaps the most difficult troubleshooting scenario is one where the system is behaving exactly as designed, but users disagree with the outcome. In these cases, the issue is not technical but communicative.

Users may expect access based on informal processes, legacy practices, or assumptions that were never formalized. The role of the architect or administrator is not to override the model impulsively, but to translate system behavior into business language and, where necessary, guide stakeholders toward a shared understanding.

Sometimes the correct resolution is not a configuration change, but a conversation.

Building Troubleshooting into the Design

The most resilient sharing models anticipate troubleshooting needs from the start. They include

- Clear RowCause definitions

- Predictable access patterns

- Intentional revocation logic

- Documentation that explains "why," not just "how"

When these elements are present, troubleshooting becomes an exercise in confirmation rather than investigation. When they are absent, even simple issues can spiral into confusion and risk.

Lessons Learned from the Field

Troubleshooting sharing issues is not a sign of failure; it is an inevitable part of operating a dynamic system that balances access, security, and compliance. The goal is not to eliminate questions, but to ensure that every answer is grounded in logic, traceable to intent, and aligned with organizational principles.

Teams that approach troubleshooting with discipline, curiosity, and respect for the underlying architecture will not only resolve issues faster—they will strengthen their sharing model over time.

In the next section, we turn our attention to how these lessons scale as organizations grow, evolve, and place increasing demands on their sharing architecture.

Scaling Your Sharing Architecture for Growth
Why Sharing Models Break at Scale

As organizations expand, the challenge is no longer about *granting* access—it is about *containing* it. Growth introduces overlapping responsibilities, temporary roles, regional variations, and exceptions that were never part of the original design. What once felt intuitive becomes opaque: access paths multiply, responsibility becomes diluted, and unintended visibility starts to surface in places no one actively designed for.

At this stage, the risk is not that users lack access, but that access accumulates silently. Records remain visible long after the original justification has expired, teams inherit permissions without understanding their origin, and administrators struggle to trace visibility back to a single, defensible rule. When a sharing model reaches this point, it does not collapse dramatically; instead, it erodes trust—users question why they can see certain data, auditors flag inconsistencies, and support teams rely on guesswork rather than certainty.

A scalable sharing architecture must therefore shift its focus from convenience to traceability. Visibility should always be attributable to a clear, current business condition, and the system should make it obvious when that condition is no longer valid. Growth exposes any model that lacks this discipline, turning minor shortcuts into long-term liabilities.

Understanding Growth Drivers That Impact Sharing

Before scaling a sharing architecture, it is essential to understand *what kind of growth* the organization is experiencing. Not all growth stresses sharing models in the same way.

> **User growth** introduces more roles, more peer-level access expectations, and greater scrutiny over "who can see what."

> **Data growth** increases query complexity and amplifies inefficiencies in recalculation and visibility evaluation.

Process growth adds new states, exceptions, and transitional access needs that were never part of the original model.

Geographic growth introduces regulatory boundaries, regional autonomy, and jurisdiction-specific visibility constraints.

A scalable sharing architecture does not attempt to flatten these differences. Instead, it **absorbs them without collapsing**.

Designing for Change, Not Perfection

One of the most important mindset shifts when scaling sharing is abandoning the idea of a "perfect" model. Instead, the goal should be **controlled adaptability**.

At scale, the question is no longer:

"Does this rule handle every case?"
It becomes

"Can this model evolve without breaking existing access guarantees?"

Sharing logic that is tightly coupled to today's org structure—specific roles, specific team names, specific field values—becomes brittle when the organization restructures. Scalable designs favor **abstractions over specifics**, even if that means slightly more upfront planning.

For example, instead of anchoring visibility to named teams or departments, mature architectures rely on *capability-based access, contextual roles,* or *functional groupings* that can outlive organizational reshuffles.

Separating Core Visibility from Contextual Access

As organizations scale, a single layer of sharing logic is rarely sufficient. High-performing architectures distinguish between

- **Core Visibility:** Access that defines who fundamentally owns or participates in a record

- **Contextual Access:** Temporary or situational visibility driven by process state, oversight needs, or collaboration windows

Blurring these two leads to confusion. Core visibility should change slowly and predictably. Contextual access should be explicit, time-aware, and auditable.

When these layers are clearly separated, scaling becomes manageable. Teams can adjust contextual access rules—such as review cycles, escalations, or special approvals—without destabilizing baseline access.

Avoiding "Visibility Inflation" Over Time

A subtle scaling risk is **visibility inflation**, where more and more users gain access to records over time without a corresponding mechanism to remove it.

This often happens when

- Temporary access is granted but never revoked.

- Exception-based visibility becomes the norm.

- Teams add new sharing paths without auditing existing ones.

At small scale, visibility inflation may go unnoticed. At enterprise scale, it becomes a compliance and trust issue.

Scalable architectures treat **revocation as a first-class concern**, not an afterthought. Every access grant should implicitly answer the question: *Under what condition should this access disappear?*

Designing for Predictable Recalculation

As record counts grow, recalculation becomes a dominant factor in sharing performance. While Salesforce handles much of this internally, architectural choices heavily influence how often and how broadly recalculations occur.

Scalable designs minimize unnecessary churn by

- Reducing dependencies on frequently changing fields

- Avoiding cascading access logic triggered by unrelated updates

- Ensuring that access decisions are driven by stable, meaningful state changes

A model that recalculates access every time a minor field changes will struggle as volumes increase. A model that ties access changes to deliberate lifecycle transitions will remain stable even under load.

Segmenting Data to Limit Blast Radius

One of the most effective scaling strategies is **intentional segmentation**. Instead of treating all records as part of a single-visibility universe, mature organizations divide data into logical partitions.

Segmentation may be based on

- Business unit or program

- Region or jurisdiction

- Record type or lifecycle stage

- Sensitivity classification

The goal is not isolation for its own sake, but **limiting the blast radius** of any sharing change. When access logic applies only to a defined segment, growth in one area does not degrade performance or clarity elsewhere.

Supporting Asynchronous Growth Patterns

As organizations scale, not all access changes happen in real time. Some are best handled asynchronously to preserve system responsiveness and user experience.

Examples include the following:

- Periodic reassessment of long-running records

- Cleanup of legacy access

- Alignment of visibility after bulk data migrations

- Enforcement of new compliance rules across historical data

A scalable sharing architecture acknowledges that **not all visibility needs are urgent**. Separating real-time access decisions from background reconciliation processes allows the system to grow without becoming brittle.

Preparing for Mergers, Restructures, and Reorgs

Growth is rarely linear. Mergers, acquisitions, and restructures place extraordinary stress on sharing models because they introduce overlapping hierarchies, duplicate roles, and conflicting visibility expectations.

Architectures that survive these events share common traits:

- Access logic is documented and explainable.

- Ownership and sharing responsibilities are clearly distinguished.

- Temporary coexistence of multiple models is supported.

- Clean rollback paths exist.

Rather than attempting to normalize everything immediately, scalable designs allow transitional states without compromising data protection.

Scaling Without Overexposing Sensitive Data

As teams grow, there is often pressure to "open up" visibility to reduce friction. While well-intentioned, this approach can quietly undermine compliance and user trust.

Scalable sharing models resist blanket access expansions. Instead, they

- Make access purposeful rather than convenient

- Prefer explicit grants over implicit inheritance

- Preserve least-privilege principles even under growth pressure

This discipline becomes especially important in regulated environments, where growth often increases scrutiny rather than reducing it.

Institutional Knowledge As a Scaling Asset

One overlooked aspect of scaling is **knowledge continuity**. As organizations grow, the people who designed the original sharing model may no longer be present.

Architectures that scale well are those that can be understood by someone new. This means

- Clear naming conventions

- Consistent patterns across objects

- Shared mental models between administrators, architects, and auditors

When a sharing model requires oral history to explain, it is already at risk.

Recognizing When to Redesign, Not Patch

Perhaps the most difficult scaling decision is knowing when incremental fixes are no longer sufficient. There comes a point where adding another exception, another condition, or another workaround introduces more risk than value.

Signs that a redesign may be necessary include the following:

- Conflicting access outcomes that are hard to explain

- Frequent emergency fixes to visibility issues

- Performance degradation tied to sharing recalculation

- Growing reliance on manual overrides

Redesigning a sharing architecture is disruptive, but at scale, **controlled disruption is often safer than unmanaged complexity**.

Scaling As a Continuous Discipline

Ultimately, scaling your sharing architecture is not a one-time initiative. It is an ongoing discipline that evolves alongside the organization itself.

The most resilient models are those treated as living systems—reviewed regularly, challenged thoughtfully, and refined deliberately. They do not aim to eliminate complexity, but to contain it in ways that remain transparent and defensible.

When sharing scales well, users trust the system. When it does not, confidence erodes quietly but steadily. The difference is rarely accidental; it is architectural.

With a scalable foundation in place, the final challenge is sustaining trust and control over time. This requires not only technical soundness, but governance, accountability, and shared ownership.

The following section provides industry insights on common pitfalls highlighting the options available for optimizing the model, how to troubleshoot, and then test your understanding through the Brainstorming.

Tips from Industry Experts on Sharing

After years of building, fixing, and re-architecting sharing models across vastly different Salesforce orgs, one truth becomes clear: **no two sharing implementations age the same way.** What works elegantly in one organization may collapse under pressure in another—not because the platform failed, but because human behavior, governance maturity, and organizational culture evolved in unexpected ways.

To ground this chapter in real-world experience rather than theory, this section draws directly from insights shared by Salesforce Certified Technical Architects and senior practitioners who have worked on some of the most complex enterprise and nonprofit implementations. Their perspectives come not from ideal scenarios, but from projects that survived audits, restructures, data migrations, regulatory reviews, and rapid scale.

Rather than presenting "best practices" as rigid rules, these experts highlight **patterns they trust**, **mistakes they avoid**, and **signals they watch for** when a sharing model begins to drift away from its original intent. Some insights reinforce platform fundamentals; others challenge commonly accepted assumptions about roles, rules, and automation.

The insights that follow come from Salesforce practitioners and architects who generously shared their real-world experiences designing, implementing, and maintaining sharing and visibility models at scale.

Each contributor was invited to reflect on the same core questions, but their responses naturally vary in structure and depth. Some chose a concise, framework-driven approach, while others shared free-flowing reflections drawn directly from years of hands-on work. To preserve the authenticity of these perspectives, their inputs are presented exactly as they were shared, without editorial modification.

Every voice included here represents a unique lens shaped by different organizations, challenges, and lessons learned. Together, they offer a grounded view of how sharing and visibility decisions play out beyond diagrams and documentation—where theory meets operational reality.

I am deeply grateful to each of them for taking the time to contribute to this chapter and, in doing so, strengthening the practical value of this book for the wider Salesforce community.

Architect Commentary: In Their Own Words

(The following insights are reproduced as shared by the contributors.)

Name: Buyan Thyagarajan

Current Company: Eigen X

Designation: Principal Architect

About Me: I am a Forbes Author/Salesforce MVP (2015-17)/
Marketing champion/Salesforce-certified architect specializing
in public sector, higher education, and manufacturing for over
15 years. I am a blogger passionate in sharing knowledge and
helping clients increase their sales using Salesforce. I am currently
working on making IT developers/Architects leverage Vibe
Architecture which will help them to be proactive, increase their
brand, and elevate their career.

What Are the Common Pitfalls You've Seen in Designing Sharing and Visibility Models in Salesforce?

I would like to list a couple of pitfalls which always cause problems down the line.

- Ignoring role hierarchy for flat organizations or implementing role
  hierarchy with titles in an organization which causes a huge mess

- Making OWD public read/write on objects which can lead to security
  nightmares

- For portal users, creating Apex sharing rules for each object which
  can lead to performance issues

What Are the Early Warning Signs That This Pitfall Might Be Happening?

- Too many sharing rules for sharing data to represent hierarchical needs like a VP, Divisional Manager, Manager, and Sales Person having sharing rules instead of hierarchy.

- Users complaining of too much data access or seeing wrong data which they should not see. Sales people complaining on why accounts are showing up which are not part of the territory. Too many duplicates also cause the problem.

- Slowness on page load times for portal users.

How Would You Recommend Overcoming This Pitfall?

- I would recommend a strategy on role hierarchy on Day1 of the implementation. If the organization is flat, have one or two standard roles and have users a part of it. If there is a hierarchical organization like a sales organization (VP of sales, Division Manager, Manager, Sales Rep), implement hierarchy based on their role, territory instead of job title.

- For OWD, try to keep objects private. For objects which would need global access, have it Public Read Only.

- If there is a need to share more than five objects for portal users based on business rule-based sharing, explore master-detail relationships which can cut down the sharing rules.

Let us say there is a claimant whose claim is submitted on their behalf by their law firm, and a public agency is reviewing the claim internally. If the business rule says share the claim to law firm users and their employees based on approval, an Apex sharing rule with a master-detail relationship can really cut down rules as the children will inherit the parent.

What Is One Piece of Advice You Would Give to Aspiring Architects Who Want to Design Better Sharing Models?

Think of least privilege access to start with, design for optimal use and think about performance and maintenance impact on any security rule implementation.

Anything Else You'd Like to Add About Scalability, Performance, or Security in Salesforce Sharing Models?

Any security solution should scale to one to million users without major changes. If your security solution will cause delays on page loads, deployments with rules running in the background, it is time to revisit them. Security is a critical model of Salesforce well-architected framework (EAT) and has to be considered for every solution.

Name: Dave Norris

Current Company: Salesforce

Designation: Developer Advocate

In this chapter we explore the world of sharing and visibility through the eyes of people who have been developing on the Salesforce platform for many years. Each person gives us their insights into common pitfalls when configuring visibility and sharing in Salesforce with tips to avoid them.

The art of architecting for visibility and sharing lies in creating scalable designs that grow with your organization while maintaining data security. Poor architectural decisions often don't surface immediately—their impact may only become apparent weeks, months, or years after implementation. Remember that the most elegant sharing model is typically the simplest one that meets the requirements. Start simple and add complexity only when necessary.

While there are numerous best practices for efficient sharing design, let's focus on three common pitfalls that can significantly impact your org's performance and maintainability.

1. Over-Reliance on Your Role Hierarchy

Many implementations attempt to mirror their company's organizational structure within Salesforce's role hierarchy. While this approach might seem logical for stakeholder communication, it often creates unnecessary complexity that impacts both maintenance and performance.

Warning Signs

- Role hierarchy exceeds ten levels deep

- Frequent role hierarchy modifications to accommodate org structure changes

- Regular user role updates due to department changes

- Performance degradation during bulk data updates

- Users experiencing record locking contentions

Better Approach

- Keep role hierarchy under ten levels—fewer is better

- Focus on logical data access groups rather than org chart structure

- Combine multiple organizational levels into single-role hierarchy levels where appropriate

- Regularly audit and eliminate unused roles

- Schedule role hierarchy changes outside business hours to minimize sharing recalculation impact

---

Pro Tip   Utilize the Salesforce Optimizer report to identify unused roles that can be removed or consolidated.

---

2. Inefficient Sharing Rules

Projects often accumulate criteria-based and owner-based sharing rules over time to address security requirements. Without proper governance, this can lead to significant maintenance and performance challenges.

Warning Signs

- More than 50 sharing rules for a single object. Anything over 30 should be audited and checked.

- Overlapping criteria-based rules.

- Slow record saves times.

- Approaching sharing rule governor limits.

- Inconsistent access patterns—users report records visible when they shouldn't be and vice versa.

Better Approach

- Begin with restrictive Organization-Wide Defaults (OWDs)

- Open access strategically through carefully planned sharing rules

- Consolidate overlapping criteria-based rules

- Implement regular sharing rule audits

- Document business justification for each rule

Pro Tip   Review the ObjectPermissions, SharingRule, and Share tables to identify consolidation opportunities. This analysis should be performed with both an Administrator and Technical Architect.

---

3. Incorrect Record Ownership Assignment Strategy

Poor ownership planning can create unintended access patterns and performance issues, particularly during bulk operations. Common anti-patterns include defaulting to admin ownership or using catch-all owner accounts.

Warning Signs

- High concentration of records owned by a single user

- System Administrator appearing as record owner

- User complaints about overly permissive or restrictive access

- Performance issues during bulk data operations

- Frequent manual sharing requirements

Better Approach

- Design clear object-level ownership strategies

- Implement automated ownership assignment processes

- Regular monitoring of ownership patterns

- Plan for scale in ownership distribution

Technical Note   Ownership skew frequently causes sharing and visibility issues, particularly record locking during updates. When changing record owners, the group membership table must be updated. Parallel data load jobs requiring group membership updates can create table locking conflicts. The more ownership changes, the more severe these issues become.

Pro Tip   Implement regular monitoring of record ownership patterns using standard reports. Look for concentration of ownership and address any skew proactively before it impacts performance. For most standard objects keep ownership under 10,000 records per user. For objects with complex sharing or are frequently updated consider a number closer to 3,000 records per user.

If a large number of configuration changes is expected consider deferring sharing calculation until after the data load job completes.

Name: Dinesh Yadav

Current Company: IBM

Designation: Senior Salesforce Architect

Ten Architect-Approved Steps to Designing Scalable Sharing and Visibility

1. Start with the Data Model, Not the Sharing Model

- Understand object relationships and record ownership before defining access

- Keep ownership meaningful—avoid system or integration users owning large datasets

- Normalize relationships so you can leverage implicit sharing (account-team, opportunity-team, etc.)

2. Layered Access: Design from Inside Out

Think of visibility as concentric rings:

- Organization-Wide Defaults (OWDs) → baseline

- Role Hierarchy → managerial visibility

- Sharing Rules → lateral/functional access

- Manual Sharing/Teams → ad hoc needs

- Apex/Criteria-based Sharing → automation or complex logic

- External/Territory/Digital Experience → specialized overlays

3. Avoid Apex Sharing If Possible

- Apex sharing should be your last resort—it's expensive and hard to maintain.

- Try criteria-based rules, territory models, or teams first.

- If you must use it, batch operations and handle share recalculations asynchronously.

4. Keep It Scalable

- Limit the number of sharing recalculations—large data volumes can cripple performance.

- Use "With Sharing" and "Without Sharing" judiciously in Apex.

- Evaluate group membership expansion and avoid role hierarchy depth > 10.

- Use Deferred Sharing Recalculation in bulk operations.

5. Design for Clarity and Governance

- Document every sharing rule and its business justification

- Create a Sharing and Visibility Decision Matrix that maps each object's access level

- Regularly audit Group Memberships, Territory Assignments, and Manual Shares

- Use Salesforce Optimizer and Health Check for proactive visibility reviews

6. Use the Right Tools

- Setup ➤ Sharing Settings for quick baseline checks

- Sharing Button/Debug Logs to trace access

- Setup Audit Trail to monitor changes to sharing config

- Enterprise Territory Management (ETM) for complex sales orgs

- External Sharing Model for Experience Cloud sites

7. Don't Forget External Users

- Distinguish between Customer, Partner, and Guest visibility models

- Use Secure Guest User Record Access (enabled by default since Spring '24)

- Always test Experience Cloud access with real profiles—not just internal roles

8. Test, Simulate, Validate

- Use "View All / Modify All" sparingly — it breaks visibility testing.

- Set up user personas to test end-to-end access.

- Document "expected vs actual" access during UAT to catch inheritance issues early.

9. Align with Compliance and Data Residency

- For regulated industries, map sharing logic to compliance zones (e.g., HIPAA, GDPR).

- Consider Shield Platform Encryption, Field-Level Security (FLS), and Transaction Security Policies.

10. Communicate the Model

- Provide visual diagrams of sharing architecture (role tree, record visibility flow).

- Train admins on how sharing rules interact—avoid accidental overexposure..

- Build a runbook for onboarding new admins to maintain consistency.

Name: Jitendra Zaa

Current Company: IBM

Designation: CTO - CRM & Service Industries

About Me: CTA, Author, 9xMVP

What Are the Common Pitfalls You've Seen in Designing Sharing and Visibility Models in Salesforce?

Security testing gets pushed to UAT when teams are already running against project timelines. Developers say "it works fine in my dev sandbox," QA says "we tested fine in QA sandbox"—but neither actually tested with the right persona because they all had elevated permissions. User stories don't define personas properly, so nobody knows what access each user type should actually have. CI/CD pipelines are not in sync across environments, meaning security metadata like profiles and permission sets drift without anyone noticing. The result? Tons of defects from basic Salesforce security—OWD, sharing rules, FLS—all discovered at the worst possible time.

What Are the Early Warning Signs That This Pitfall Might Be Happening?

User stories say "user can access records" without specifying which user type or persona. Dev and QA sandboxes don't have test users configured for each persona—everyone just logs in as admin or with elevated permissions. When developers confirm something works, nobody asks "tested as which user?" There's no security design document before development starts. And the biggest red flag: security defects start flooding in during UAT, all related to "can't see this record" or "shouldn't be able to see that field."

How Would You Recommend Overcoming This Pitfall?

Security by Design from Day 1—define OWD, sharing rules, and FLS during solution design, not during UAT. Every user story must specify the persona and what they can and cannot access. Create persona-based test users in all sandboxes from the start—Dev, QA, UAT—with permissions that match production. Never test with System Administrator. Keep CI/CD pipelines in sync so security metadata doesn't drift between environments. It doesn't take rocket science—just apply basic first principles and treat security as a design decision, not an afterthought.

What Is One Piece of Advice You Would Give to Aspiring Architects Who Want to Design Better Sharing Models?

Nobody talks about security until something goes wrong—and by then it's the most expensive mistake you'll ever make. Almost every data breach in Salesforce implementations happens not because Salesforce isn't secure, but because it wasn't designed properly. In an age where data is currency, one oversight in your sharing model can expose customer data, violate compliance, and bankrupt a company. Take security seriously from Day 1. It's not a checkbox at the end—it's the foundation you build everything on.

Anything Else You'd Like to Add About Scalability, Performance, or Security in Salesforce Sharing Models?

All the best practices are available in Salesforce documentation—anyone can read them. What makes an architect different from an admin or developer is the ability to apply that knowledge in real projects by navigating pushbacks from stakeholders, project managers, and non-Salesforce enterprise architects. An architect's job is to educate clients, ARBs, and leadership on why security by design and performance testing matter—even when they can't see business value on a user screen or a shiny new capability for the business. The hard part isn't knowing the patterns; it's convincing everyone else to invest time in things that only matter when they go wrong.

Name: Lilith Van Biesen

Current Company: Capgemini

Designation: Principal Architect, Director

About Me: Lilith Van Biesen (she/hers) is a Principal Architect and Director at Capgemini, a Salesforce CTA and MVP. She started working with Salesforce in 2017 as a "reskiller" and has loved the platform and its credentials ever since. She is a frequent public speaker at Salesforce and Salesforce community events focusing on topics of architecture, women in tech, and the value of diversity and inclusion.

What Are the Common Pitfalls You've Seen in Designing Sharing and Visibility Models in Salesforce?

A common pitfall I have seen in sharing architecture is the overcomplication of solutions. Typical root causes can be an overbloated business process without a clear strategy supporting the requirements or an insufficient understanding of the various sharing mechanisms in the Salesforce toolset.

What Are the Early Warning Signs That This Pitfall Might Be Happening?

If a process or solution has more exceptions than compliance, cannot be captured into simple rules or cannot be subdivided into modular elements, there might be more work to be done in terms of discovery and design.

How Would You Recommend Overcoming This Pitfall?

An architect should ensure there is an in-depth understanding of business processes and a thorough knowledge of the platform sharing options along with their benefits and limitations. Another mitigation is getting a second opinion or a second pair of eyes on your solution. If you are unable to explain the why and how of your solution easily, there likely is more work to be done.

Finally, architects should keep the 80–20 rule in mind and challenge to what extent the 20% should dominate the design.

What Is One Piece of Advice You Would Give to Aspiring Architects Who Want to Design Better Sharing Models?

Aside from making sure they understand the processes and tooling, they should work on understanding what their stakeholders care about and how designs fulfil their needs. This means speaking their language/jargon and can use this in providing the bigger picture or impact of a particular design. For example, a shiny custom solution might be accommodating the 20%, but may result in a significantly higher total cost of ownership (TCO). It's important for an architect to challenge the possibly redundant complexity and provide the full picture in terms stakeholders understand.

Anything Else You'd Like to Add About Scalability, Performance, or Security in Salesforce Sharing Models?

The KISS principle ("Keep it Simple, Stupid!") remains relevant in all parts of solution design. While not all problems will have a simple and elegant solution, complex solutions should have architects do a double-take and verify that it is truly the best possible option.

Name: Matt Francis

Current Company: IBM

Designation: CTA, CTO, DE

About Me: I am Salesforce Industries and AI CTO at IBM Americas, CTA, DE. My objective is to bring leading edge AI POV, Solutions and capabilities to all companies.

My other goals and objectives are as follows:

- Moving AI into everything we do so we can focus on the more important parts of life and work

- Learning and evolving daily

- Making more meaningful connections with people

What Are the Common Pitfalls You've Seen in Designing Sharing and Visibility Models in Salesforce?

I think the real challenge is organizational, if sharing can be done across groups, that is the best model, when you limit based on role, records, or some custom model, things get very complicated and thinking about all the different ways you can share can be a real challenge. The other big area is Experience Cloud and how that interacts with the sharing model, this is a serious pitfall for a lot of implementations and requires a lot of thought and consideration.

What Are the Early Warning Signs That This Pitfall Might Be Happening?

Organizational complexity.

How Would You Recommend Overcoming This Pitfall?

Working to see how business units can work together vs. siloed.

What Is One Piece of Advice You Would Give to Aspiring Architects Who Want to Design Better Sharing Models?

Never create an APEX sharing model (or code-based model)—leverage OOTB.

Name: Walter Spinrad

Current Company: Crowe

Designation: Managing Director

About Me: I am a 30+ IT veteran connected to the Salesforce Ecosystem since 2008. Salesforce Alumni and a CTA. I focus on guiding customers to successful Salesforce implementations on their path to digital transformation.

What Are the Common Pitfalls You've Seen in Designing Sharing and Visibility Models in Salesforce?

Overexposed data.

What Are the Early Warning Signs That This Pitfall Might Be Happening?

Testing exposes the risk of too much data being visible for users.

How Would You Recommend Overcoming This Pitfall?

Identify the company data policies and align sharing rules and Organization-Wide Defaults to comply.

What Is One Piece of Advice You Would Give to Aspiring Architects Who Want to Design Better Sharing Models?

Adhere to the principle of least privilege when exposing data.

Anything Else You'd Like to Add About Scalability, Performance, or Security in Salesforce Sharing Models?

Use criteria-based sharing and restrictive sharing sparingly. Avoid data skew.

Name: Anonymous Contributor

Designation: General Manager and Certified Technical Architect (CTA)

About Me: 42x Salesforce Certified, 10 years in Salesforce, 8 years in Communications Cloud/CPQ, Certified Technical Architect (CTA), Dreamforce Speaker

What Are the Common Pitfalls You've Seen in Designing Sharing and Visibility Models in Salesforce?

Many Salesforce architects and developers do not understand sharing in general, and especially Experience Cloud sharing which can leverage sharing sets and grant access to records related to Account/Contact via Account Contact Relationships. This leads to either going with Customer Community Plus or higher licenses unnecessarily (increasing license cost and impacting performance/scalability of number of customers that can be on the platform), or mis-designing the data model as a workaround.

What Are the Early Warning Signs That This Pitfall Might Be Happening?

Opting for Customer Community Plus instead of Customer Community licenses, lack of use of sharing sets

How Would You Recommend Overcoming This Pitfall?

A draft data model should be reviewed with a senior architect that understanding sharing in detail before moving to build to avoid rework.

What Is One Piece of Advice You Would Give to Aspiring Architects Who Want to Design Better Sharing Models?

Understand all the sharing mechanisms at your disposal—Organization-Wide Defaults, sharing rules, role hierarchy, sharing sets, etc.

Brainstorming

Rethinking Sharing Through an Architect's Lens

By this stage in the chapter, you have explored how sharing models behave under stress—during audits, restructures, data growth, and organizational change. This final section is not about learning new mechanisms. It is about slowing down and questioning assumptions that may have gone unchallenged since your sharing model was first introduced.

The purpose of this brainstorming exercise is not to "fix" your sharing architecture immediately, but to help you *see it clearly*. Mature sharing models are rarely the result of clever logic alone; they are the outcome of repeated reflection, course correction, and intentional simplification over time.

Use the prompts in this section as discussion starters—with your team, your stakeholders, or even just with yourself during an architecture review.

1. Re-examining the Original Intent

Every sharing model begins with a story—often one that is no longer told.
Take a moment to reflect on the early days of your implementation

- What business concern triggered the need for this sharing model?

- Who was most vocal about access at that time?

- What risks were being actively managed, and which were merely assumed?

Now Contrast That with the Present:

- Do those original drivers still exist?

- Has the organization outgrown the assumptions baked into the design?

- Are there access rules today that no one can confidently explain?

Reflection Prompt:
If you had to re-justify your current sharing model to a new leadership team tomorrow, which parts would feel solid—and which would feel uncomfortable to defend?

2. Mapping Access to Real Business Roles (Not Job Titles)

Over time, sharing logic often drifts away from real responsibilities and starts aligning instead with system artefacts—profiles, roles, permission sets, or public groups whose names no longer reflect reality.
Pause and ask

- Does access align with *what people actually do,* or with how the org chart looked years ago?

- Are temporary responsibilities (covering roles, acting managers, short-term projects) handled intentionally—or ignored entirely?

- Do exceptions feel like edge cases, or have they quietly become the norm?

Reflection Prompt:

If roles, profiles, and public groups were wiped clean tomorrow, how would you describe access purely in terms of responsibilities and outcomes?

3. Understanding Why Records Are Shared—Not Just with Whom

One of the most fragile points in mature orgs is not excessive access, but untraceable access.

Consider the following

- Can you explain *why* a specific user can see a specific record without querying multiple layers of configuration?

- Are reasons for access documented anywhere outside of code or setup?

- Would an auditor or new architect be able to follow the trail?

Reflection Prompt:

Pick a randomly shared record in your org. How many steps does it take to understand why access exists—and how confident are you in that explanation?

4. Visualizing the Lifecycle of Access

Access is often granted decisively but removed hesitantly.

Reflect on access lifecycles

- When a user's role changes, does access automatically adapt—or accumulate?

- Are there records that should no longer be visible but remain accessible "just in case"?

- Is access removal treated as a first-class concern, or an afterthought?

Reflection Prompt:

Which access paths in your org have a clear end condition—and which ones rely on manual cleanup or hope?

5. Stress-Testing for Organizational Change

Most sharing models are designed for stability, not transformation.

Imagine scenarios such as

- A merger that doubles user count overnight

- A regional restructure that dissolves existing role hierarchies

- A regulatory change requiring stricter data segmentation

- A new executive mandate for cross-team transparency

Reflection Prompt:

Which of these scenarios would require a redesign of your sharing model—and which could it absorb without breaking?

6. Identifying Silent Performance Risks

Performance issues tied to sharing rarely announce themselves clearly. They surface as

- Slow list views

- Inconsistent reports

- Queries that behave differently across users

- Timeouts that appear unrelated to access logic

Reflect carefully

- Do performance investigations routinely rule out sharing logic without evidence?

- Are there areas where access logic is "too complex to touch"?

- Have performance trade-offs been consciously accepted—or accidentally inherited?

Reflection Prompt:

If you had to remove 20% of your sharing logic tomorrow for performance reasons, which parts would you hesitate to touch—and why?

7. Balancing Transparency and Control

Every organization claims to value transparency—until access becomes uncomfortable. Consider the following

- Where has sharing been restricted due to political sensitivity rather than genuine risk?

- Are there records that remain private because "that's how it's always been"?

- Conversely, where has access been widened without clear accountability?

Reflection Prompt:

Which access decisions in your org are driven by trust—and which by fear?

8. Evaluating Governance Maturity

Sharing models reflect governance maturity more than technical skill. Ask yourself

- Is there a clear ownership model for access decisions?

- Who is responsible when sharing logic becomes outdated?

- Are changes reviewed architecturally—or handled reactively?

Reflection Prompt:

If a critical access issue emerged today, would responsibility be clear—or would it become a collective scramble?

9. Listening to User Behavior

Users often reveal sharing flaws without realizing it. Observe the following

- Frequent requests for "temporary" access

- Screenshots shared instead of record access

- Shadow processes outside Salesforce

- Users unsure whether they *should* have access

Reflection Prompt:

What user behaviors in your org might be compensating for weaknesses in your sharing model?

10. Designing for the Next Five Years, Not the Last Five

Finally, shift your perspective forward.

Instead of asking

"What does our sharing model protect today?"

Ask

- What kind of organization are we becoming?

- What types of collaboration will increase?

- Where will regulatory pressure intensify?

- Which access decisions will be hardest to undo later?

Reflection Prompt:

If you were designing this sharing model for an organization twice your current size, what would you deliberately simplify?

11. Questioning Invisible Power Structures

Not all access decisions are technical; some are social.

Over time, informal power dynamics can quietly shape who sees what, even when those decisions are no longer consciously defended.

Consider whether certain teams or individuals have retained broader access simply because they were early adopters, long-tenured, or historically influential.

Reflection Prompt:

If access were reassessed without names attached—only responsibilities—whose visibility would change?

12. Separating Convenience from Necessity

Many sharing decisions are made to reduce friction rather than meet true business needs. What starts as convenience can harden into permanent access.

Pause to examine where access exists primarily to "avoid questions," "save time," or "reduce tickets," rather than to enable accountable work.

Reflection Prompt:

Which access paths exist mainly to reduce short-term inconvenience—and what risks do they quietly introduce?

13. Identifying Access That Exists Only Because It Once Did

Some access remains simply because removing it feels risky, not because it is still justified.

This inertia can be difficult to challenge, especially when no incident has occurred to force a review.

Reflection Prompt:

Which areas of your sharing model survive solely because "nothing bad has happened yet"?

14. Considering the Cost of Over-Protection

While excessive access is often discussed, excessive restriction has its own cost—delayed decisions, fragmented context, and duplicated work.Think about where over-protection might be slowing collaboration or forcing users into workarounds.

Reflection Prompt:

Where might tighter sharing controls be solving the wrong problem?

15. Examining How New Hires Experience Visibility

New users often experience the sharing model most honestly.

Their confusion, hesitation, or reliance on others to navigate access can expose misalignment between design and reality.

Reflection Prompt:

If a new hire had to explain your sharing model after 30 days, what gaps would they highlight?

16. Reviewing Emergency Access Patterns

Temporary access granted during crises—deadlines, outages, escalations—often becomes permanent by default.

These moments reveal how flexible or brittle the sharing model really is.

Reflection Prompt:

What emergency access granted in the past still exists today—and was it ever intentionally reviewed?

17. Challenging the Assumption That More Rules Mean More Safety

As organizations mature, sharing logic often grows denser, not clearer.

More rules can sometimes obscure responsibility rather than enforce it.

Reflection Prompt:

If you removed half of your explicit sharing rules, would accountability become clearer—or more chaotic?

18. Understanding Where Trust Replaces Design

In some areas, access exists not because it was designed, but because teams trust each other to "do the right thing."

Trust can be powerful—but fragile when teams change.

Reflection Prompt:

Where does your sharing model rely on personal trust rather than structural clarity?

19. Observing How Often Sharing Is Discussed—And How

The tone of conversations about access matters.

Frequent tension, defensiveness, or avoidance around sharing decisions can signal deeper architectural misalignment.

Reflection Prompt:

When sharing comes up in meetings, does it feel like problem-solving—or damage control?

20. Defining What "Good Enough" Looks Like

Perfection in sharing is neither realistic nor sustainable.

Mature architectures are not flawless; they are intentionally bounded.

Reflection Prompt:

If you had to define a "good enough" sharing model for your organization, what compromises would you consciously accept—and which would you refuse?

Closing Thoughts

Sharing and visibility are often treated as background mechanics—something that either works or doesn't. In reality, they are quiet indicators of how an organization thinks about trust, accountability, and growth. Every access decision encodes an assumption about people: who needs context, who needs protection, and who is expected to act responsibly with information.

As your Salesforce org evolves, the technical correctness of a sharing model matters far less than its clarity. Models that endure are not the most complex or the most restrictive; they are the ones that can be explained plainly, adjusted deliberately, and defended with confidence when questioned. They leave behind a trail of intent, not just configuration.

This chapter has asked you to slow down—to observe how access behaves over time, how users respond to it, and how small compromises quietly accumulate. That pause is not a luxury. It is how architects prevent tomorrow's urgency from being shaped by yesterday's shortcuts.

If there is one principle worth carrying forward, it is this: a healthy sharing model is never "finished." It is reviewed, challenged, and refined as the organization changes. When visibility feels predictable, auditable, and aligned with real work—not just system structure—you are no longer managing access. You are designing trust into the platform.

CHAPTER 7

Case Studies and Sample Exam Questions for Salesforce Sharing and Visibility

The case studies and sample exam questions in this chapter are designed to allow you to transition directly from concept mastery into a focused assessment experience. This setup will assist those focusing on comprehensive exam prep.

Why This Chapter Exists

By the time architects and senior practitioners reach the end of a book on sharing and visibility, they rarely need another explanation of what Organization-Wide Defaults are or how role hierarchies behave. What they often need is something more difficult to find—a way to test their instincts.

Sharing and visibility concepts tend to feel obvious—until they are examined under pressure. Certification exams exploit this gap deliberately. They do not test whether you remember features; they test whether you understand *trade-offs*, *side effects*, and *failure modes*. Real projects do the same. This chapter is designed to serve both purposes.

It brings together carefully constructed questions, scenarios, and case studies that mirror how sharing decisions unfold in real Salesforce environments—messy, constrained, and rarely ideal. Some questions are intentionally uncomfortable. Others may appear simple until you slow down and examine what is *not* being said.

© Sandhya Sharma 2026

S. Sharma, *Unlocking Salesforce Data*, https://doi.org/10.1007/979-8-8688-2305-3_7

If earlier chapters helped you design and critique sharing models, this chapter helps you pressure-test them.

How to Use This Chapter

You do not need to read this chapter linearly.

- If you are preparing for an exam, focus on the multiple-choice and scenario sections, but read the explanations even when you answer correctly.

- If you are an architect reviewing an existing org, spend more time with the case studies, where context matters more than correctness.

- If you are mentoring others, use these questions as discussion prompts rather than quizzes.

There is no single "right" way to approach sharing. There *are*, however, many ways to be confidently wrong.

Multiple Choice Questions on Sharing Settings

The questions in this section resemble certification-style prompts, but they are written to reflect how sharing behaves in live environments—not how it is described in isolation.

Most of the questions are followed by a brief explanation that focuses on *why* the correct option holds, not just *what* it is.

1. **What determines the baseline level of record access in Salesforce?**

   A. Sharing Rules

   B. Profiles

   C. Organization-Wide Defaults

   D. Role Hierarchy

   Answer: C

   Explanation:

   Organization-Wide Defaults (OWDs) define the most restrictive baseline access before any other sharing mechanisms are applied.

2. **If an object's OWD is set to Private, who can see a record by default?**

   A. All users

   B. Only the record owner

   C. Users in the same role

   D. Users with Edit permission

   Answer: B

   Explanation:

   With Private OWD, only the record owner can see the record by default. Access for others depends on role hierarchy, sharing rules, or manual/Apex sharing.

3. **What happens when "Grant Access Using Hierarchies" is disabled on a custom object?**

   A. Sharing rules stop working.

   B. Manual sharing is removed.

   C. Role hierarchy access is ignored.

   D. Profiles lose object access.

   Answer: C

   Explanation:

   Disabling this setting prevents users higher in the role hierarchy from automatically gaining access to records.

4. **Which sharing mechanism is evaluated first?**

   A. Manual sharing

   B. Apex-managed sharing

   C. Organization-Wide Defaults

   D. Sharing Rules

   Answer: C

Explanation:

OWD establishes the baseline level of access before additional sharing layers are evaluated.

5. **What access does a user gain when they are above the record owner in the role hierarchy?**

   A. Read only

   B. Edit

   C. Same as owner

   D. Depends on OWD and object permissions

   Answer: D

   Explanation:

   Role hierarchy grants access based on OWD and object permissions.

6. **Which statement about role hierarchies is true?**

   A. They replace sharing rules.

   B. They control record ownership.

   C. They grant access vertically.

   D. They work only for standard objects.

   Answer: C

   Explanation:

   Role hierarchies grant access upward through management levels.

7. **Can role hierarchies restrict access?**

   A. Yes

   B. No

   C. Only for custom objects

   D. Only with permission sets

   Answer: B

Explanation:

Role hierarchies only open access; they never restrict it.

8. **What type of sharing rule uses record ownership as criteria?**

   A. Criteria-based

   B. Apex-managed

   C. Owner-based

   D. Manual

   Answer: C

   Explanation:

   Owner-based sharing rules share records owned by specific users, roles, or public groups.

9. **Which field types can be used in criteria-based sharing rules?**

   A. Formula fields

   B. Picklists, checkboxes, numbers, dates

   C. Lookup fields

   D. Long text areas

   Answer: B

   Explanation:

   Criteria-based sharing supports specific field types such as picklists, checkboxes, numbers, and dates.

10. **Why can't sharing rules reference related objects?**

   A. Performance limitations

   B. Data security concerns

   C. Salesforce design constraints

   D. Licensing restrictions

   Answer: C

Explanation:

Sharing rules evaluate only fields on the record itself, not related objects.

11. **What access level can be granted via sharing rules?**

A. Read only

B. Read/Write

C. Read, Write, Delete

D. Full access

Answer: B

Explanation:

Sharing rules can grant Read or Read/Write access only.

12. **What happens when a record no longer meets a sharing rule's criteria?**

A. Access remains.

B. Access is revoked automatically.

C. Record ownership changes.

D. Manual cleanup is required.

Answer: B

Explanation:

Salesforce automatically removes access granted by sharing rules when criteria are no longer met.

13. **Manual sharing allows a user to:**

A. Change OWD

B. Grant access to individual users

C. Modify role hierarchy

D. Create sharing rules

Answer: B

Explanation:

Manual sharing allows ad hoc access to individual users or groups.

14. **Who can manually share a record?**

A. Any user

B. System Administrator only

C. Record owner or users with Full Access

D. Users with Read access

Answer: C

Explanation:

Only record owners or users with Full Access permission can manually share records.

15. **Which feature supports temporary access without automation?**

A. Sharing rules

B. Permission sets

C. Manual sharing

D. Profiles

Answer: C

Explanation:

Manual sharing is ideal for one-off or temporary access needs.

16. **Public groups are best used to**

A. Replace roles

B. Simplify sharing assignments

C. Assign licenses

D. Enforce OWD

Answer: B

Explanation:

Public groups simplify sharing by grouping users logically.

17. **What happens if a user is removed from a public group?**

    A. Their role changes.

    B. Their profile resets.

    C. Shared access is revoked.

    D. Record ownership transfers.

    Answer: C

    Explanation:

    Access granted through public groups is removed after sharing recalculation.

18. **Which object type always enforces role hierarchy?**

    A. Most standard objects

    B. Custom objects

    C. External objects

    D. All objects

    Answer: A

    Explanation:

    Most standard objects always enforce role hierarchy and do not allow disabling it.

19. **Can permission sets grant record-level access?**

    A. Yes

    B. No

    C. Only with Apex

    D. Only for custom objects

    Answer: B

    Explanation:

    Permission sets control object and field access, not record visibility.

20. **What is the primary risk of excessive sharing rules?**

    A. Data loss

    B. Performance degradation

    C. Licensing violations

    D. Profile conflicts

    Answer: B

    Explanation:

    Complex sharing logic can negatively impact query and recalculation performance.

21. **Which sharing option is best for predictable, stable access?**

    A. Manual sharing

    B. Apex-managed sharing

    C. Sharing rules

    D. Temporary permissions

    Answer: C

    Explanation:

    Declarative sharing rules are easier to maintain and predictable.

22. **What happens if OWD is set to Public Read/Write?**

    A. Sharing rules are ignored.

    B. All users can read and edit records.

    C. Manual sharing is disabled.

    D. Roles stop working.

    Answer: B

    Explanation:

    Public Read/Write allows all users to read and edit records, subject to object permissions.

23. **Why is "Public Read Only" often preferred over "Public Read/Write"?**

    A. It improves reporting.

    B. It reduces security risk.

    C. It simplifies profiles.

    D. It disables sharing rules.

    Answer: B

    Explanation:

    It limits edit access while still enabling visibility.

24. **Which sharing layer is the most granular?**

    A. OWD

    B. Role hierarchy

    C. Sharing rules

    D. Manual sharing

    Answer: D

    Explanation:

    Manual sharing operates at individual record and user level.

25. **What determines whether a user can delete a record?**

    A. Sharing settings

    B. Profile or permission set

    C. Role hierarchy

    D. OWD

    Answer: B

    Explanation:

    Delete permissions are controlled by object permissions.

26. **What determines whether a standard object has a share object available?**

A. Whether the object is customizable.

B. Whether the object supports record-level sharing.

C. Whether the object has lookup relationships.

D. Whether the object is included in reports.

Answer: B

Explanation:

Only objects that support record-level sharing (private or controlled OWD) have corresponding share objects.

27. **In which scenario is Apex-managed sharing the *only* viable option?**

A. Sharing records based on record owner's role

B. Granting access when a picklist equals a fixed value

C. Sharing records based on a calculated condition spanning multiple related objects

D. Granting access to a public group

Answer: C

Explanation:

Declarative sharing rules cannot reference fields on related objects or evaluate complex, cross-object conditions. Apex-managed sharing is required when visibility depends on logic spanning multiple objects or derived conditions that cannot be expressed declaratively.

28. **What happens when multiple sharing rules apply?**

A. The most restrictive wins.

B. The least restrictive wins.

C. Salesforce throws an error.

D. Only the first rule applies.

Answer: B

Explanation:

Salesforce grants the highest level of access available.

29. **Can a record be shared via both rule and Apex?**

    A. No

    B. Yes

    C. Only for standard objects

    D. Only temporarily

    Answer: B

    Explanation:

    Multiple sharing mechanisms can coexist.

30. **What is the safest default OWD for sensitive data?**

    A. Public Read/Write

    B. Public Read Only

    C. Private

    D. Controlled by Parent

    Answer: C

    Explanation:

    Private ensures access is explicitly granted.

31. **Controlled by Parent means**

    A. Child inherits parent access.

    B. Parent inherits child access.

    C. Role hierarchy is ignored.

    D. Sharing rules don't apply.

    Answer: A

    Explanation:

    Child records inherit access from their parent.

32. **Which object commonly uses Controlled by Parent?**

   A. Account

   B. Contact

   C. Opportunity

   D. Custom junction objects

Answer: D

Explanation:

Junction objects typically inherit access from parents.

33. **What happens if a parent record becomes private?**

   A. Child records remain visible.

   B. Child records inherit restriction.

   C. Only owners retain access.

   D. Sharing rules override it.

Answer: B

Explanation:

Child access follows the parent's sharing model.

34. **What does "implicit sharing" refer to?**

   A. Sharing via Apex

   B. Sharing via criteria rules

   C. System-generated access

   D. Manual sharing

Answer: C

Explanation:

Salesforce automatically grants access through certain relationships.

35.   **Which relationship creates implicit sharing?**

A. Lookup

B. Master-detail

C. External lookup

D. Hierarchical

Answer: B

36.   **Which access level is never granted via sharing?**

A. Read

B. Edit

C. Delete

D. Full Access

Answer: C

37.   **Why is the RowCause field critical in Apex-managed sharing?**

A. It determines whether sharing is manual or automatic.

B. It controls record ownership.

C. It distinguishes programmatic shares from other access paths.

D. It improves query performance automatically.

Answer: C

Explanation:

RowCause identifies *why* access exists and allows architects to safely manage, audit, and revoke programmatic shares without interfering with owner-, rule-, or manual-based sharing entries.

38.   **What should be reviewed before adding a new sharing rule?**

A. Profile permissions

B. Existing sharing logic

C. Role hierarchy

D. All of the above

Answer: D

39. **Which feature is best for cross-team visibility without ownership change?**

    A. Ownership reassignment

    B. Public groups

    C. Profiles

    D. Data categories

    Answer: B

40. **What happens if OWD is Private and no sharing exists?**

    A. All users can view.

    B. Only admins can view.

    C. Only owners can view.

    D. Records become hidden.

    Answer: C

41. **What is the primary architectural risk of using "Manual" RowCause in Apex sharing?**

    A. Salesforce restricts bulk inserts.

    B. Manual shares cannot be deleted.

    C. Programmatic and user-created shares become indistinguishable.

    D. It increases governor limit consumption.

    Answer: C

    Explanation:

    Using the Manual RowCause blurs the line between user intent and automation. This makes audits, troubleshooting, and safe revocation extremely difficult in mature orgs.

42. **Which statement about share objects is TRUE?**

   A. They are queried only when sharing rules change.

   B. They are evaluated during every record-level access check.

   C. They exist only for custom objects.

   D. They are cached per user session.

Answer: B

Explanation:

Salesforce consults share tables during access evaluation for every query, list view, and report. Poorly designed sharing logic directly impacts system performance at scale.

43. **Why should revocation logic be treated as a first-class design concern?**

   A. Salesforce automatically cleans up obsolete shares.

   B. Revocation reduces SOQL usage.

   C. Access conditions change more frequently than access grants.

   D. Revocation improves UI responsiveness.

Answer: C

Explanation:

Access requirements evolve constantly due to role changes, restructures, and lifecycle events. Without deliberate revocation logic, access accumulates silently, increasing compliance and security risk.

44. **What architectural smell indicates a sharing model has drifted?**

   A. Too many sharing rules

   B. Inability to explain *why* a user has access

   C. High record ownership changes

   D. Frequent role hierarchy updates

Answer: B

Explanation:

When access cannot be clearly justified, the model has likely grown organically without governance, making it fragile, risky, and hard to audit.

45. **Which sharing mechanism evaluates *after* object-level security but *before* Apex logic?**

    A. Permission Sets

    B. Role Hierarchy

    C. Sharing Rules

    D. Apex-managed sharing

    Answer: C

    Explanation:

    Sharing rules extend record-level access after object-level security is enforced, and this access is evaluated before any Apex logic runs.

46. **Why should triggers not directly contain sharing logic?**

    A. Triggers cannot insert share records.

    B. Sharing logic in triggers causes deployment failures.

    C. Triggers lack transactional control.

    D. It tightly couples access logic to data mutation events.

    Answer: D

    Explanation:

    Embedding sharing logic directly in triggers makes it hard to test, reuse, scale, and reason about. Service-layer isolation preserves architectural clarity and long-term maintainability.

47. **What is the most reliable indicator that a sharing model will not scale?**

    A. High user count

    B. Deep role hierarchy

    C. Access logic based on assumptions instead of ownership

    D. Large data volumes alone

    Answer: C

    Explanation:

    Scaling issues usually stem from fragile assumptions about behavior and responsibility, not from volume alone. Growth exposes these assumptions quickly.

48. **Which approach best supports audit readiness?**

    A. Minimizing the number of share records

    B. Centralizing all access through one role

    C. Using descriptive custom RowCauses

    D. Relying on implicit sharing via ownership

    Answer: C

    Explanation:

    Custom RowCauses provide explicit reasoning for access decisions, allowing auditors and architects to trace, validate, and justify visibility paths.

49. **What is the primary downside of excessive public group usage?**

    A. Slower UI rendering

    B. Increased deployment complexity

    C. Loss of contextual intent behind access

    D. Governor limit exhaustion

    Answer: C

246

Explanation:

Public groups often become abstract containers disconnected from real business responsibilities, making access difficult to explain or govern over time.

50. **Which scenario most strongly argues against role hierarchy usage?**

    A. Linear management structures

    B. Matrix organizations with temporary teams

    C. Small sales teams

    D. Simple ownership models

    Answer: B

    Explanation:

    Role hierarchies assume stable reporting lines. Matrix and project-based organizations require dynamic, context-driven access models.

51. **Why is "too complex to touch" a dangerous phrase in sharing design?**

    A. It indicates lack of documentation.

    B. It prevents performance optimization.

    C. It signals unmanaged risk concentration.

    D. It increases admin dependency.

    Answer: C

    Explanation:

    Complexity avoidance often hides accumulated risk. Untouched sharing logic becomes brittle, untestable, and increasingly dangerous during change.

52.   **Which user behavior most commonly signals a visibility problem?**

A. High login frequency

B. Increased report usage

C. Screenshots shared instead of record links

D. Reduced page load times

Answer: C

Explanation:

When users resort to screenshots, they are compensating for missing or inconsistent access—a clear signal of sharing gaps.

53.   **Why should sharing logic be reviewed during reorganizations?**

A. Salesforce requires redeployment.

B. Ownership changes automatically reset sharing.

C. Access assumptions no longer align with responsibilities.

D. Record counts increase.

Answer: C

Explanation:

Reorgs change *who does what,* not just titles. Sharing models tied to outdated assumptions quickly become incorrect.

54.   **What is the safest way to support time-bound access?**

A. Manual sharing

B. Role-based access

C. Scheduled revocation logic

D. Profile cloning

Answer: C

Explanation:

Time-bound access requires automated removal. Manual approaches almost always leave residual access behind.

55. **Which design principle most improves long-term sharing stability?**

    A. Maximum automation

    B. Minimal configuration

    C. Explicit intent documentation

    D. Centralized admin control

    Answer: C

    Explanation:

    Clear intent prevents drift, simplifies audits, and allows future architects to understand *why* decisions were made.

56. **Why is over-sharing often harder to detect than under-sharing?**

    A. Users rarely report excess access.

    B. Reports hide shared records.

    C. Salesforce masks extra visibility.

    D. Logs do not capture it.

    Answer: A

    Explanation:

    Users complain when blocked, not when over-privileged. Over-sharing often persists unnoticed until audits or incidents occur.

57. **What should architects prioritize during sharing refactors?**

    A. Reducing share record counts

    B. Preserving behavioral intent

    C. Matching legacy behavior exactly

    D. Minimizing test coverage

    Answer: B

Explanation:

Refactors should protect *why* access exists, not blindly replicate outdated implementations.

58.   **Why should sharing logic be resilient to partial failure?**

A. Salesforce enforces it.

B. Batch jobs always fail partially.

C. Large data volumes increase error probability.

D. UI operations depend on it.

Answer: C

Explanation:

At scale, failures are inevitable. Architected sharing logic must handle partial success gracefully without corrupting access state.

59.   **What does "security by design" mean in sharing architecture?**

A. Adding controls after incidents

B. Locking down all data

C. Designing access intentionally from inception

D. Relying on profiles

Answer: C

Explanation:

Security by design embeds access decisions into architecture early, rather than reacting to problems later.

60.   **Which metric is *least* useful when evaluating sharing health?**

A. Number of share records

B. User access requests

C. Ability to explain access paths

D. Volume of rejected access attempts

Answer: A

Explanation:

Raw counts alone lack context. Understanding intent and clarity matters far more than absolute numbers.

61. **Why is ownership-based access insufficient in complex orgs?**

    A. Ownership is slow.

    B. Records change owners frequently.

    C. Responsibilities extend beyond ownership.

    D. Salesforce limits ownership.

    Answer: C

    Explanation:

    Modern collaboration models require visibility that transcends simple ownership structures.

62. **What architectural choice most reduces future refactoring?**

    A. Deep role hierarchies

    B. Descriptive naming conventions

    C. Over-granting access

    D. Manual approvals

    Answer: B

    Explanation:

    Clear naming preserves intent and prevents misunderstanding as teams evolve.

63. **What distinguishes architect-level sharing decisions from admin-level ones?**

    A. Use of Apex

    B. Focus on compliance and scale

C. Advanced UI configuration

D. Certification level

Answer: B

Explanation:

Architects design for longevity, governance, compliance, and organizational change—not just immediate functionality.

64. **Why is Apex-managed sharing considered a governance risk if poorly designed?**

A. It bypasses profiles.

B. It cannot be audited.

C. It creates access paths outside declarative visibility controls.

D. It disables role hierarchy.

Answer: C

Explanation:

Apex-managed sharing introduces programmatic access paths that are not immediately visible in standard setup screens, making governance, audits, and long-term ownership harder if not documented and controlled.

65. **Which factor most strongly influences query performance when complex sharing is enabled?**

A. Number of users

B. Size of role hierarchy

C. Volume of share table rows

D. Record type count

Answer: C

Explanation:

Salesforce evaluates share tables during query execution. Excessive or fragmented share rows significantly increase query cost, especially in large-data-volume orgs.

66. **Why should architects avoid relying on manual sharing in enterprise-scale orgs?**

    A. Manual sharing is deprecated.

    B. It cannot be reported on.

    C. It depends on user behavior rather than governed logic.

    D. It only supports read access.

    Answer: C

    Explanation:

    Manual sharing introduces inconsistency and human dependency, making access unpredictable and difficult to audit or scale across teams and geographies.

67. **What is the primary architectural concern with deeply nested role hierarchies?**

    A. Licensing cost

    B. Profile conflicts

    C. Increased sharing recalculation complexity

    D. UI navigation issues

    Answer: C

    Explanation:

    Deep hierarchies multiply implicit access paths, increasing recalculation time and making visibility outcomes harder to reason about as the organization grows.

68. **Which scenario most clearly requires custom RowCause usage?**

    A. Owner-based access

    B. Temporary project-based collaboration

    C. Territory reassignment

    D. Profile-driven access

    Answer: B

Explanation:

Custom RowCauses provide traceability for nonstandard, business-driven access that must be clearly distinguishable from system-generated sharing.

69. **Why is revocation logic more critical than grant logic at scale?**

    A. Grants are irreversible.

    B. Revocations trigger recalculations.

    C. Access accumulation creates compliance risks.

    D. Revocations consume more limits.

Answer: C

Explanation:

Unrevoked access silently accumulates over time, creating security and compliance risks that are often discovered only during audits or incidents.

70. **What architectural signal suggests a sharing model has drifted from its intent?**

    A. Increased use of permission sets

    B. Frequent "temporary" access requests

    C. Higher login frequency

    D. Reduced automation

Answer: B

Explanation:

Repeated temporary access requests often indicate that the underlying sharing logic no longer reflects real business responsibilities.

71. **Which design principle best supports long-term sharing maintainability?**

    A. Maximum automation

    B. Centralized service-layer ownership

    C. Minimal documentation

    D. Profile consolidation

    Answer: B

    Explanation:

    Centralized ownership ensures consistent logic, controlled changes, and accountability as org complexity increases.

72. **Why are sharing-related performance issues often misdiagnosed?**

    A. Salesforce hides errors.

    B. Logs do not capture sharing.

    C. Symptoms appear indirect and inconsistent.

    D. Sharing logic runs asynchronously.

    Answer: C

    Explanation:

    Sharing issues manifest as slow queries, inconsistent list views, or user-specific behavior, making root cause analysis nonobvious.

73. **Which governance practice most effectively reduces accidental over-sharing?**

    A. Quarterly role reviews

    B. Annual security audits

    C. Explicit access ownership definitions

    D. Increased admin permissions

    Answer: C

Explanation:

Clear ownership ensures someone is accountable for why access exists and when it should be reviewed or removed.

74. **What is the architectural risk of using public groups as long-term access containers?**

    A. Group limits

    B. UI clutter

    C. Membership drift over time

    D. Reduced flexibility

    Answer: C

    Explanation:

    Public groups tend to accumulate members without regular review, making access justification increasingly opaque.

75. **Which question best tests sharing model resilience?**

    A. Can admins explain it?

    B. Can it survive organizational restructuring?

    C. Does it meet current requirements?

    D. Is it fully automated?

    Answer: B

    Explanation:

    Architectural resilience is measured by how well a model adapts to change without requiring complete redesign.

76. **Why should architects document "why" access exists, not just "how"?**

    A. For faster development

    B. For training purposes

C. For auditability and future decision-making

D. For UI clarity

Answer: C

Explanation:

Understanding intent prevents accidental removal, overextension, or duplication of access logic during future changes.

77. **What distinguishes sharing logic designed for growth versus stability?**

A. More rules

B. Fewer users

C. Built-in adaptability to change

D. Higher automation

Answer: C

Explanation:

Growth-oriented designs anticipate restructuring, scale, and evolving compliance needs rather than optimizing only for the current state.

78. **Which user behavior most often indicates hidden sharing flaws?**

A. High login frequency

B. Frequent report exports

C. Reliance on screenshots instead of record access

D. Use of mobile app

Answer: C

Explanation:

When users share data outside Salesforce, it often signals access gaps or mistrust in the visibility model.

79. **Why should architects treat sharing logic as a living system?**

A. Salesforce updates frequently.

B. Business roles evolve continuously.

C. Licenses change yearly.

D. UI features improve.

Answer: B

Explanation:

Sharing reflects organizational behavior, which changes over time, making static designs brittle.

80. **Which factor most complicates sharing audits in mature orgs?**

A. Lack of reports

B. Excessive record types

C. Multiple overlapping access paths

D. UI customization

Answer: C

Explanation:

Overlapping paths obscure the true reason for access, complicating validation and remediation.

81. **Why should performance testing include multiple user personas?**

A. To test profiles

B. To validate role hierarchy

C. To expose sharing-dependent query differences

D. To test UI rendering

Answer: C

Explanation:

Different access paths result in different query plans, making persona-based testing essential.

82. **What architectural mistake most often leads to emergency sharing fixes?**

    A. Under-automation

    B. Over-documentation

    C. Reactive access decisions

    D. Limited testing

    Answer: C

    Explanation:

    Reactive decisions bypass design review, creating fragile shortcuts that accumulate technical debt.

83. **Which practice best prevents silent access sprawl?**

    A. Limiting admin count

    B. Scheduled access reviews

    C. Reducing automation

    D. Locking profiles

    Answer: B

    Explanation:

    Regular reviews surface obsolete access before it becomes a compliance issue.

84. **Why is simplicity a strategic goal in sharing architecture?**

    A. It reduces setup time.

    B. It improves UI performance.

    C. It lowers cognitive and operational risk.

    D. It eliminates Apex.

    Answer: C

Explanation:

Simpler models are easier to explain, govern, audit, and adapt over time.

85. **What should trigger a formal sharing architecture review?**

    A. New Salesforce release

    B. Increased user complaints

    C. Organizational restructuring

    D. UI redesign

    Answer: C

    Explanation:

    Structural change alters access assumptions, making prior designs potentially invalid.

86. **Which design choice most improves explainability of access?**

    A. Fewer profiles

    B. Explicit RowCause definitions

    C. More automation

    D. Reduced record types

    Answer: B

    Explanation:

    Clear RowCauses make access intent traceable for architects, auditors, and future teams.

87. **Why should sharing models be reviewed during mergers or acquisitions?**

    A. License alignment

    B. Profile conflicts

    C. Differing data ownership expectations

    D. UI consistency

    Answer: C

Explanation:

Merging organizations often have incompatible assumptions about data visibility and control.

88. **Which mindset best supports long-term sharing success?**

   A. Feature-driven

   B. Rule-driven

   C. Behavior-driven

   D. Tool-driven

Answer: C

Explanation:

Sharing exists to support how people work, not just how systems are configured.

89. **What is the ultimate goal of an architect-designed sharing model?**

   A. Maximum restriction

   B. Minimal configuration

   C. Predictable, justified, and scalable access

   D. Complete automation

Answer: C

Explanation:

Architect-level sharing balances security, usability, governance, and growth without relying on fragile assumptions.

90. **When should an architect explicitly avoid criteria-based sharing rules?**

   A. When record ownership changes frequently.

   B. When access depends on related object state.

   C. When OWD is Public Read/Write.

   D. When reports must be optimized.

Answer: B

Explanation:

Criteria-based sharing cannot evaluate conditions across related records. Architect-level designs must avoid declarative sharing when access depends on external or parent object state.

91. **What is the primary architectural risk of excessive public groups in sharing?**

   A. Increased setup complexity

   B. UI clutter

   C. Exponential recalculation and audit opacity

   D. Higher license consumption

Answer: C

Explanation:

At scale, public groups multiply recalculation paths and make access lineage difficult to trace—both critical architectural concerns.

92. **Which sharing construct is most resilient during organizational restructuring?**

   A. Role hierarchy

   B. Manual sharing

   C. Responsibility-driven logic

   D. Criteria-based rules

Answer: C

Explanation:

Architects design for responsibility and outcomes rather than reporting lines, which are the first to change during restructures.

93. **Why do architects prefer explicit revocation logic over reliance on recalculation?**

    A. Recalculation is asynchronous.

    B. Recalculation skips inactive users.

    C. Recalculation does not guarantee access removal.

    D. Recalculation ignores Apex sharing.

    Answer: C

    Explanation:

    Implicit revocation creates access drift. Architect-grade models always define how and when access is actively removed.

94. **What is the strongest indicator that a sharing model lacks governance maturity?**

    A. Too many profiles

    B. Absence of RowCause documentation

    C. Overuse of permission sets

    D. Minimal Apex usage

    Answer: B

    Explanation:

    If access reasons cannot be explained or audited, governance has already broken—regardless of technical correctness.

95. **Which scenario most justifies time-bound sharing logic?**

    A. Internal role promotions

    B. Data migrations

C. External partner collaboration

D. Sandbox refreshes

Answer: C

Explanation:

External access must always include expiry logic. Architects treat indefinite external visibility as a compliance risk.

96. **What should an architect review first when sharing-related performance degrades?**

A. Apex CPU time

B. Trigger recursion

C. Volume of implicit share rows

D. Profile assignments

Answer: C

Explanation:

Query performance is directly impacted by the size and complexity of evaluated share tables—not just code execution.

97. **Why is "temporary access" a red flag in mature orgs?**

A. It increases admin workload.

B. It bypasses permission sets.

C. It often becomes permanent without intent.

D. It cannot be reported on.

Answer: C

Explanation:

Architects recognize that most long-term exposure begins as "temporary" access that was never revoked.

98. **What distinguishes architect-level sharing audits from admin-level reviews?**

    A. Use of tooling

    B. Depth of documentation

    C. Focus on intent, lineage, and risk

    D. Frequency of review

    Answer: C

    Explanation:

    Architects audit *why* access exists and *what happens if it shouldn't.*

99. **Which access decision is hardest to reverse once data volume scales?**

    A. Manual shares

    B. Role hierarchy changes

    C. Wide OWD relaxation

    D. Apex-managed sharing

    Answer: C

    Explanation:

    Relaxing OWD often cascades into irreversible transparency assumptions and downstream dependencies.

100. **What mindset best defines an architect designing sharing and visibility?**

    A. Security-first

    B. Performance-first

    C. Change-first

    D. User-first

    Answer: C

Explanation:

Architects design sharing models that survive change—users, scale, regulation, and structure—not just today's requirements.

Scenario-Based Questions on Visibility and Permissions

1. A global manufacturing company operates with private OWD for Opportunities. Regional sales managers must see opportunities owned by sales reps in their region, but global leadership must not automatically inherit access unless explicitly required.

   Question: How should the architect design this visibility model?

   Model Answer: Use role hierarchy only for regional layers and exclude global leadership from automatic inheritance. Provide leadership access through criteria-based sharing rules or Apex-managed sharing to ensure visibility is intentional, not implicit.

2. A nonprofit organization uses Projects and Grants. Projects should only be visible to users once the Grant funding source is confirmed as "External." Funding details live on a related Grant object.

   Question: What is the most appropriate approach to enforce this visibility?

   Model Answer: Use Apex-managed sharing, as standard sharing rules cannot evaluate fields on related objects. The architect should design logic that evaluates the Grant record and grants or revokes access accordingly.

3. A regulated healthcare org requires that users retain record access even if their role changes temporarily, but access must be revoked once the assignment officially ends.

   Question: Which architectural principle should guide this design?

Model Answer: Access lifecycle management. The architect should implement time-bound sharing using Apex or scheduled jobs, ensuring access has a defined expiry independent of role hierarchy.

4. An enterprise has heavily customized profiles controlling object access. Over time, visibility issues have become unpredictable.

   Question: What architectural correction should be applied?

   Model Answer: Decouple object access from visibility by simplifying profiles and moving record access logic to permission sets and sharing mechanisms. Architects should avoid overloading profiles with visibility intent.

5. A user reports seeing records they should not. Investigation shows access originates from multiple overlapping sharing rules.

   Question: What should the architect address first?

   Model Answer: Access traceability. The architect should rationalize sharing rules, document RowCauses where applicable, and reduce overlapping access paths to ensure explainable visibility.

6. A consulting firm frequently onboards contractors who need short-term read access to client cases.

   Question: Which visibility approach best balances security and maintainability?

   Model Answer: Use public groups combined with Apex-managed or scheduled sharing tied to contract dates, ensuring automated revocation without manual intervention.

7. After a merger, two Salesforce orgs are consolidated. Each has different role hierarchies and sharing logic.

   Question: What should guide the new unified sharing design?

   Model Answer: Business responsibility alignment, not legacy hierarchy. Architects should redesign sharing based on future operating models rather than merging old structures.

8. A report behaves differently for two users with identical profiles and permission sets.

    Question: What is the most likely cause?

    Model Answer: Record-level sharing differences. Architects must consider implicit sharing, team membership, manual shares, and Apex-managed access when diagnosing report inconsistencies.

9. A financial services org wants auditors to review records without affecting operational visibility or hierarchy.

    Question: What is the correct architectural pattern?

    Model Answer: Create auditor-specific access via permission sets and read-only sharing rules or Apex-managed sharing, avoiding role hierarchy changes.

10. An org uses Account Teams extensively, but access is inconsistent across Opportunity and Case objects.

    Question: What architectural principle is being violated?

    Model Answer: Consistency of access propagation. Architects should ensure team-based access behaves predictably across related objects or document intentional differences.

11. A global support team needs visibility only to cases escalated beyond a certain severity.

    Question: Which sharing approach fits best?

    Model Answer: Criteria-based sharing on Case severity, avoiding role hierarchy expansion that would grant unnecessary access.

12. An org wants managers to see subordinates' records only when acting as approvers.

    Question: What should the architect recommend?

    Model Answer: Use process-driven, conditional sharing rather than permanent hierarchy-based access.

13. A platform event updates records asynchronously, triggering unexpected access grants.

Question: What architectural safeguard is missing?

Model Answer: Context-aware sharing logic. Architects must design event-driven sharing with idempotency and revocation controls.

14. Users request visibility to historical records from projects they no longer support.

Question: How should the architect respond?

Model Answer: Assess business justification. Historical access should be explicit and policy-driven, not inherited by default.

15. A role hierarchy exceeds recommended depth and causes maintenance issues.

Question: What corrective action should be taken?

Model Answer: Flatten the hierarchy and shift access logic to public groups and sharing rules. Architects should avoid using hierarchy as a catch-all.

16. A new compliance rule requires segmentation by country, but existing sharing is region-based.

Question: What is the correct architectural response?

Model Answer: Refactor sharing logic to align with regulatory boundaries, even if it means re-architecting role and group structures.

17. A user retains access after being removed from a public group.

Question: What should the architect investigate?

Model Answer: Other access paths such as manual shares, teams, Apex-managed sharing, or implicit parent access.

18. A high-volume org experiences slow list views after adding new sharing logic.

    Question: What is the likely cause?

    Model Answer: Excessive or complex sharing rules increasing sharing recalculation cost. Architects must balance access granularity with performance.

19. Leadership wants visibility into all records "just in case."

    Question: How should an architect handle this request?

    Model Answer: Challenge the requirement and propose audit-based or exception-driven access rather than blanket visibility.

20. A sandbox refresh breaks sharing logic validation.

    Question: What architectural gap does this reveal?

    Model Answer: Lack of automated tests and documentation around sharing behavior. Architects should treat sharing as testable logic.

21. An org relies heavily on manual sharing for exceptions.

    Question: What long-term risk does this pose?

    Model Answer: Untraceable access accumulation. Architects should replace manual sharing with governed mechanisms wherever possible.

22. Users see records via "View All" permission unexpectedly.

    Question: What architectural lesson applies?

    Model Answer: System permissions override sharing logic. Architects must separate operational admin access from business visibility.

23. A new product line introduces different confidentiality rules.

    Question: How should sharing adapt?

    Model Answer: Introduce record type-aware sharing strategies rather than modifying existing global rules.

24. An org needs temporary cross-team collaboration without permanent access changes.

    Question: What is the best approach?

    Model Answer: Time-bound public groups or Apex-managed sharing with defined removal conditions.

25. A user requests edit access but only needs visibility.

    Question: What architectural principle applies?

    Model Answer: Least privilege. Architects must distinguish visibility from edit rights.

26. Reports used for executive decisions exclude records unexpectedly.
    Question: What should the architect verify?
    Model Answer: Whether reports run in user context or with "Run as System," and how sharing impacts reporting outcomes.

27. An org migrates from Classic to Lightning and visibility issues surface.

    Question: Why does this happen?

    Model Answer: Lightning surfaces sharing inconsistencies more clearly through components and dynamic pages.

28. Users can see child records but not parents.

    Question: What architectural misalignment exists?

    Model Answer: Sharing logic is not aligned across object relationships. Architects must consider implicit sharing directions.

29. A department requests access exceptions weekly.

    Question: What does this indicate?

    Model Answer: The base sharing model no longer matches operational reality and requires redesign.

30. A partner community user sees internal records unintentionally.

Question: What should be reviewed first?

Model Answer: Sharing sets.

31. A user can access records via API but not UI.

Question: What should be checked?

Model Answer: Object permissions vs. record-level access and field-level security differences.

32. A business unit wants autonomy over access rules.

Question: What governance approach fits best?

Model Answer: Centralized architectural control with configurable metadata-driven sharing logic.

33. A new org is designed with Public Read/Write OWD for speed.

Question: What risk does this introduce?

Model Answer: Security debt. Architects should design for controlled visibility even if initial speed is a priority.

34. An org disables "Grant Access Using Hierarchies."

Question: When is this appropriate?

Model Answer: When hierarchy inheritance conflicts with strict compliance or separation-of-duties requirements.

35. Users complain about unpredictable access after automation changes.

Question: What should architects enforce?

Model Answer: Change management discipline for sharing-impacting automation.

36. A record owner leaves the organization.

Question: What should happen to access?

Model Answer: Ownership transfer should not implicitly widen or restrict access without review.

37.   A team relies on reports instead of record access.

Question: What does this signal?

Model Answer: Sharing may be too restrictive or poorly aligned with business workflows.

38.   A compliance audit questions why a user had access six months ago.

Question: What capability is missing?

Model Answer: Access justification and historical traceability.

39.   A department insists on exceptions "for convenience."

Question: How should architects respond?

Model Answer: Evaluate long-term governance impact before accommodating convenience-based access.

40.   A single object has over 20 sharing rules.

Question: What risk does this pose?

Model Answer: Maintenance and performance degradation. Architects should consolidate logic.

41.   A new Salesforce release changes sharing recalculation behavior.

Question: What should architects do?

Model Answer: Review release notes and test sharing-heavy processes proactively.

42.   Users see records via search but not list views.

Question: What explains this?

Model Answer: Search indexing timing combined with sharing recalculation lag.

43.   An org introduces delegated administration.

Question: What must architects ensure?

Model Answer: Delegated admins cannot unintentionally bypass sharing governance.

44. A business unit bypasses Salesforce for sensitive data.

    Question: What does this reveal?

    Model Answer: Lack of trust in the sharing model's effectiveness.

45. A new integration user needs broad access.

    Question: How should this be handled?

    Model Answer: Integration-specific permission sets and controlled sharing, not system admin access.

46. Access rules differ across environments.

    Question: What is the risk?

    Model Answer: Production behavior becomes unpredictable after deployments.

47. Users ask, "Should I be seeing this?"

    Question: What does this indicate?

    Model Answer: Poor visibility explainability.

48. A record appears in dashboards but not reports.

    Question: Why?

    Model Answer: Dashboard running user context differs from report viewer context.

49. A new law restricts data residency.

    Question: What must change?

    Model Answer: Sharing logic must enforce jurisdictional boundaries.

50. An architect inherits a legacy org with undocumented sharing.

    Question: What is the first step?

    Model Answer: Audit and map existing access paths before introducing any changes.

Short Answer Questions

1. In a mature Salesforce org, why should record visibility decisions be treated as part of data architecture rather than user management?

   Answer: Record visibility directly affects data integrity, compliance boundaries, reporting accuracy, and system performance. When treated merely as user management, sharing logic becomes reactive and fragmented. Architects must view visibility as a structural concern that defines how data flows across organizational units, survives scale, and remains defensible under audit.

2. Why is relying solely on role hierarchy risky in large, multi-region implementations?

   Answer: Role hierarchy assumes stable managerial relationships, which often break down across regions, matrix teams, and temporary assignments. Overreliance leads to overexposure, deep hierarchies that hurt performance, and brittle designs that fail during restructures. Architects must design access that reflects responsibility, not reporting lines alone.

3. When should an architect discourage the use of criteria-based sharing rules?

   Answer: Criteria-based rules should be avoided when access depends on related records, time-bound conditions, or complex business states. In such cases, declarative rules become opaque and difficult to reverse. Architects must recognize when simplicity turns into hidden complexity and recommend programmatic or hybrid approaches instead.

4. How does unmanaged Apex-managed sharing create long-term operational risk?

Answer: Without clear RowCause governance and revocation strategies, Apex-managed sharing accumulates orphaned access paths. Over time, this leads to unexplained visibility, performance degradation, and audit failures. The risk lies not in Apex itself, but in treating it as a one-time solution rather than a lifecycle-managed capability.

5. Why is access revocation considered harder than access grant in enterprise orgs?

   Answer: Access grant is usually tied to a clear business event, while revocation depends on detecting when that event is no longer valid. Enterprise orgs often lack reliable signals for role changes, project completion, or temporary assignments ending. Architects must design explicit end conditions rather than relying on manual cleanup.

6. What architectural signal suggests that a sharing model has exceeded its intended complexity?

   Answer: When no single person can confidently explain why a specific user sees a specific record without investigating multiple layers of configuration, the model has crossed its maintainability threshold. Complexity becomes a liability once explainability is lost.

7. Why should public groups be governed more strictly than permission sets?

   Answer: Public groups directly influence record visibility and can aggregate access silently across many users. Unlike permission sets, their impact is often indirect and harder to trace. Architects must treat public groups as access boundaries, not convenience containers.

8. How does excessive manual sharing undermine governance?

   Answer: Manual sharing bypasses architectural intent and creates access paths that are difficult to audit or reverse at scale. Over time, it introduces inconsistency and weakens trust in the system's security posture. Architects typically restrict manual sharing to exceptional, well-documented cases.

9. Why is "with sharing" insufficient as a security guarantee in Apex design?

    Answer: "With sharing" enforces existing access rules but does not validate whether those rules are correct, current, or appropriate. Architects must ensure the underlying sharing model is sound; otherwise, code simply enforces flawed assumptions at scale.

10. How does sharing design influence report credibility?

    Answer: Reports execute in the context of user visibility. If sharing logic is inconsistent or overly complex, reports yield fragmented or misleading insights. Architects must ensure that reporting audiences align cleanly with visibility boundaries to preserve decision-making integrity.

11. Why should architects avoid designing sharing logic around individual users?

    Answer: User-centric designs do not scale and collapse during turnover. Architecture should target roles, responsibilities, or groups that represent enduring business functions. Users are transient; access principles should not be.

12. What makes RowCause strategy an architectural concern rather than a coding detail?

    Answer: RowCause defines why access exists. Without a clear strategy, share records become indistinguishable, making audits, troubleshooting, and cleanup impractical. Architects use RowCause as a semantic layer that documents intent within the system itself.

13. How does data volume amplify sharing inefficiencies?

    Answer: Every additional share record increases query complexity and memory usage. Inefficiencies that are invisible at low volume become performance bottlenecks at scale. Architects must evaluate sharing impact not per record, but across millions of access checks.

14. Why is "least privilege" difficult to enforce in fast-growing organizations?

    Answer: Rapid growth introduces exceptions faster than governance can adapt. Teams prioritize access speed over precision, leading to privilege creep. Architects must design guardrails that slow uncontrolled expansion without blocking legitimate collaboration.

15. When does simplifying a sharing model become a strategic decision?

    Answer: Simplification becomes strategic when complexity threatens performance, auditability, or organizational trust. Removing access paths can be more valuable than adding new ones. Architects recognize that subtraction is often a sign of maturity.

16. How can sandbox testing misrepresent real sharing behavior?

    Answer: Sandboxes often lack realistic data volumes, role depth, and user diversity. As a result, sharing logic appears performant and predictable when it may fail under production conditions. Architects compensate by stress-testing assumptions, not environments alone.

17. Why should sharing reviews be scheduled independently of feature releases?

    Answer: Sharing drift occurs through cumulative changes, not single deployments. Regular reviews allow architects to detect misalignment early, rather than reacting after incidents or audits expose issues.

18. What distinguishes compliance-driven visibility from collaboration-driven visibility?

    Answer: Compliance-driven visibility prioritizes restriction, traceability, and justification. Collaboration-driven visibility prioritizes openness and speed. Architects balance these forces explicitly rather than allowing one to dominate by default.

19. Why do mergers often invalidate existing sharing assumptions?

Answer: Mergers introduce overlapping roles, conflicting data ownership models, and new regulatory expectations. Sharing logic designed for a single organizational identity rarely survives integration without re-architecture.

20. How does user trust influence sharing design success?

Answer: If users do not trust that access is intentional and fair, they bypass the system through exports, screenshots, or shadow tools. Architects must design visibility that users understand, not just tolerate.

21. Why is "temporary access" a recurring architectural challenge?

Answer: Temporary access often lacks clear start and end signals, making automation difficult. Without explicit lifecycle design, temporary access becomes permanent by default. Architects must treat temporariness as a first-class requirement.

22. What role does documentation play in sustainable sharing architecture?

Answer: Documentation preserves intent beyond individual contributors. Without it, future teams cannot distinguish design decisions from accidental configurations. Architects document not just how access works, but why it exists.

23. How can performance tuning expose deeper sharing flaws?

Answer: Performance optimization often reveals unnecessary joins, excessive share records, or redundant logic. These symptoms usually point to architectural issues rather than tuning opportunities alone.

24. Why should architects question inherited sharing models during new implementations?

Answer: Inherited models reflect past constraints and assumptions. Reusing them without scrutiny risks importing obsolete logic into new contexts. Architects reassess fit rather than assuming prior correctness.

25.   What is the ultimate measure of a well-designed sharing model?

Answer: A well-designed sharing model is one that remains understandable, defensible, and adaptable years after its creation—even as the organization changes. Longevity, not cleverness, is the true benchmark.

Case Studies: Solving Near Real-World Sharing Challenges

Case Study 1: Global Nonprofit Expanding Beyond Its Original Operating Model

Organizational Context

A global nonprofit organization operates programs across education, healthcare, and disaster relief. Salesforce was originally implemented when the organization operated in only three countries, with a simple regional structure and fewer than 300 users.

The original sharing model relied heavily on role hierarchy, with Organization-Wide Defaults set to Private for Projects and Cases. Regional managers sat above local teams, and visibility flowed naturally through the hierarchy.

How the Problem Emerged

Over five years, the nonprofit expanded into 14 additional countries. New regional hubs were introduced, matrix reporting became common, and global oversight teams were formed for compliance and donor reporting.

As the organization scaled, users began reporting inconsistent visibility:

- Global teams could not see active projects in some regions.

- Regional teams accidentally gained access to projects outside their remit.

- Donor reporting teams requested manual access repeatedly.

Administrators attempted to patch these gaps using additional public groups and criteria-based sharing rules.

Why the Original Model Failed

The original role hierarchy assumed a strict vertical reporting structure. As matrix teams and cross-regional collaboration increased, role-based visibility stopped reflecting how work was actually performed.

Sharing rules multiplied, but they overlapped in ways that were difficult to explain or audit.

Architect's Evaluation

An architect reviewing this situation would ask:

- Is visibility driven by geography, program type, funding source, or all three?

- Which access is permanent vs. situational?

- Which teams require read-only visibility vs. edit access?

They would also identify that role hierarchy is no longer the primary driver of access.

Architectural Resolution

The architect restructures visibility around **program ownership and funding accountability**, not geography alone. Role hierarchy is flattened where possible, while Apex-managed sharing is introduced for donor-based and oversight access.

Access revocation is explicitly defined when projects close or funding cycles end.

Key Lesson

Scaling nonprofits rarely fail due to volume alone. They fail when sharing models remain tied to organizational charts instead of operational reality.

Case Study 2: Financial Services Firm Facing Regulatory Audit Findings

Organizational Context

A regional financial services firm uses Salesforce for loan processing and customer servicing. The org supports retail banking, SME lending, and wealth management.

Due to regulatory requirements, customer data must be segmented strictly between business units.

How the Problem Emerged

During an external audit, regulators identified that some relationship managers could see customer records outside their licensed business unit. While no misuse occurred, the firm was flagged for insufficient access controls.

Administrators believed their sharing model was compliant because

- OWD was set to Private.

- Profiles were tightly controlled.

- Manual sharing was restricted.

Root Cause

The issue stemmed from legacy sharing rules created during a previous reorganization. These rules granted visibility based on account region, not business unit, unintentionally crossing regulatory boundaries.

Architect's Evaluation

The architect focuses on

- Understanding regulatory definitions of "access"

- Mapping data visibility to legal accountability

- Identifying all implicit access paths (reports, dashboards, queues)

Architectural Resolution

The architect replaces region-based sharing with **business-unit-driven access**, enforced through a combination of record types, ownership constraints, and Apex-managed sharing for compliance teams.

A periodic audit report is introduced to surface anomalous access.

Key Lesson

Compliance failures often arise not from missing controls, but from outdated assumptions embedded in sharing logic.

Case Study 3: SaaS Company with Rapid Team Turnover

Organizational Context

A fast-growing SaaS company uses Salesforce to manage customer success, renewals, and expansion opportunities. Internal mobility is high, with employees frequently rotating between teams, regions, and acquired business units.

Salesforce access was originally designed around stable account ownership and long-term team assignments.

How the Problem Emerged

Over time, former account managers retained access to customer records long after moving to unrelated roles. Some accessed sensitive renewal forecasts unintentionally, while others appeared in reports they no longer needed to see.

Manual access cleanup became a recurring operational burden, dependent on HR notifications and administrative follow-up.

Architect's Evaluation

The architect identified a fundamental mismatch between how access was granted and how work actually flowed. Roles were temporary, but access was permanent.

The key question became

- What event should *end* access, not just grant it?

Architectural Resolution

Visibility was redesigned around active engagement rather than job title. Access became tied to assignment records with explicit start and end dates. When an engagement closed, visibility expired automatically without manual intervention.

Key Lesson

When access lacks a defined end condition, it inevitably outlives its purpose and creates silent risk.

Case Study 4: Healthcare Provider Balancing Privacy and Care Coordination

Organizational Context

A healthcare provider operates multiple clinics and outreach programs. Salesforce supports patient referrals, care coordination, and follow-up activities across organizational boundaries.

Regulatory obligations require strict protection of sensitive health information, while operational realities demand collaboration.

How the Problem Emerged

Care coordinators needed visibility into patient journeys across clinics, but privacy rules restricted full record access. Teams either received too much access or none at all, slowing care delivery.

Architect's Evaluation

The architect reframed the problem by separating *visibility* from *authority*. Not every user who needs awareness requires full clinical access.

Architectural Resolution

The solution layered access

- Non-sensitive coordination data was shared broadly.

- Sensitive clinical details remained protected through field-level controls.

- Contextual sharing ensured access aligned with care involvement.

Key Lesson

Visibility does not have to be binary. Thoughtful separation of data layers enables compliance without blocking collaboration.

Case Study 5: University Managing Cross-Department Research Grants

Organizational Context

A large university manages research grants involving faculty, departments, and external collaborators. Salesforce tracks grant lifecycles, funding milestones, and reporting obligations.

How the Problem Emerged

Visibility varied depending on who created the record. Some departments felt locked out, while others saw more than they should.

Architect's Evaluation

The architect recognized that ownership patterns were inconsistent and not meaningful in a collaborative research environment.

Architectural Resolution

Sharing was aligned to grant lifecycle stages rather than ownership. Visibility expanded and contracted as grants moved from proposal to active research to closure.

Key Lesson

Lifecycle-aware sharing is more resilient than ownership-based sharing in collaborative environments.

Case Study 6: Manufacturing Firm with Distributor Access Needs

Organizational Context

A manufacturing firm uses Salesforce to manage orders, warranties, and service requests. External distributors are granted limited access to support after-sales processes, while internal teams handle pricing, contracts, and escalation workflows.

How the Problem Emerged

As distributor access expanded, certain users began inheriting visibility into internal pricing and contract-related records. The exposure was inconsistent and difficult to reproduce, making it hard to identify the root cause quickly.

Architect's Evaluation

The architect reviewed all access paths impacting distributor users, focusing on indirect visibility through ownership, public groups, and inherited sharing rules. The evaluation highlighted that external access was interwoven with internal sharing logic instead of being explicitly isolated.

Architectural Resolution

Distributor visibility was redesigned as a self-contained access model, using dedicated groups and explicit sharing mechanisms with clear access reasons. Internal access paths were decoupled to prevent accidental inheritance.

Key Lesson

External users magnify the consequences of even small sharing misconfigurations. External access must never rely on internal sharing assumptions; it requires deliberate isolation and explicit intent.

Case Study 7: Government Agency Under FOIA Pressure

Organizational Context

A government agency uses Salesforce to manage citizen cases, correspondence, and internal notes. Data access must align with public accountability standards and Freedom of Information regulations.

How the Problem Emerged

During FOIA reviews, inconsistencies surfaced in who could view internal case notes. While technically permitted by the system, the access patterns were difficult to justify in policy terms.

Architect's Evaluation

The architect examined access not only from a technical standpoint, but through the lens of defensibility. The key question became whether each access path could be clearly explained to auditors and legal reviewers.

Architectural Resolution

Sharing logic was realigned to mirror policy language rather than system convenience. Each access path was documented, justified, and traceable to an explicit governance decision.

Key Lesson

In public-sector orgs, access must be explainable to nontechnical audiences.

Case Study 8: Retail Chain with Seasonal Workforce

Organizational Context

A global retail chain hires large numbers of seasonal staff during peak sales periods. Salesforce supports store operations, customer cases, and inventory-related workflows.

How the Problem Emerged

Temporary staff accounts often retained access after contracts ended. Manual deactivation lagged behind business timelines, creating audit and security concerns.

Architect's Evaluation

The architect identified that access was tied to user existence rather than active engagement. There was no systematic link between business timelines and visibility rules.

Architectural Resolution

Access was redesigned to align with time-bound engagement records. Visibility automatically expired when seasonal assignments ended, eliminating reliance on manual cleanup.

Key Lesson

Seasonal workforces demand access models that expire by design, not by exception.

Case Study 9: NGO Managing Sensitive Beneficiary Data

Organizational Context

An NGO manages beneficiary records across multiple programs, some of which involve highly sensitive personal information. Field teams and central teams have different access needs.

How the Problem Emerged

Field teams required operational visibility, but unrestricted access risked exposing personal identifiers. Role hierarchy changes further complicated consistency.

Architect's Evaluation

The architect focused on separating operational visibility from sensitive data exposure. Reliance on role hierarchy was identified as unstable given frequent organizational changes.

Architectural Resolution

Visibility was layered based on record state and business ownership. Sensitive attributes were protected independently of record access, ensuring privacy without blocking delivery.

Key Lesson

Ethical access design requires separating what users need to act from what they do not need to know.

Case Study 10: Enterprise CRM After Multiple Mergers

Organizational Context

An enterprise org merges three Salesforce instances. A large enterprise consolidated three Salesforce orgs following a series of mergers. Each legacy org had its own sharing philosophy and governance standards.

How the Problem Emerged

Conflicting visibility rules created confusion, inconsistent access, and compliance concerns. Attempts to preserve existing rules increased complexity rather than reducing it.

Architect's Evaluation

The architect assessed access patterns against regulatory obligations and operational usability. Legacy logic was treated as input, not as a constraint.

Architectural Resolution

A unified sharing model was rebuilt from first principles, with clearly segmented access paths and documented rationale for each decision.

Key Lesson

Mergers require intentional re-architecture; inherited sharing logic rarely survives consolidation intact.

Tips for Passing the Salesforce Security and Sharing Exam

The Salesforce Security and Sharing exam is not designed to reward memorization. Candidates who approach it as a checklist of features—roles, sharing rules, profiles, permission sets—often leave the exam surprised by how little surface-level knowledge helps under pressure. This exam tests how well you think, not how much you recall.

At its core, the exam evaluates whether you can reason about access the way an architect would: deliberately, defensibly, and with long-term consequences in mind.

Think in Terms of Intent, Not Mechanism

A recurring mistake candidates make is jumping too quickly to *how* something is implemented without first understanding *why* access exists. Exam questions are often framed to test whether you can identify the correct design intent before selecting a technical approach.

When reviewing a scenario, pause before looking at the answer options. Ask yourself

- What business risk is being managed?

- Who owns the data, and who merely consumes it?

- Is access permanent, conditional, or temporary?

Once the intent is clear, the technical solution usually becomes obvious—and distractor options become easier to eliminate.

Assume the Exam Context Is Imperfect by Design

Many scenarios intentionally omit information that would be available in a real project. This is not a flaw—it is the test.

Instead of looking for a "perfect" solution, look for the *most defensible* one:

- Which option introduces the least long-term maintenance?

- Which approach scales without constant reconfiguration?

- Which solution would survive an audit, a restructure, or leadership change?

Architect-level questions rarely ask what *can* be done. They ask what *should* be done.

Treat Over-Engineering As a Red Flag

If an option feels impressive but fragile, it is usually wrong.

The exam consistently favors

- Declarative solutions over programmatic ones

- Clear ownership over clever logic

- Predictability over flexibility

Apex-managed sharing, complex automation, or deeply nested access logic should only be selected when the scenario explicitly demands it. If simpler mechanisms can satisfy the requirement, the exam expects you to choose them.

Read for What Changes, Not What Exists

Many questions hinge on *change*—a new region, a merger, regulatory expansion, or user role transitions. The correct answer is often the one that handles change gracefully rather than optimizing for the current state.

Ask yourself

- What happens when this condition no longer applies?

- How easily can access be revoked?

- Who will understand this logic two years from now?

Solutions that acknowledge lifecycle—not just initial access—are almost always preferred.

Watch for Implicit Governance Signals

Architect-level questions often include subtle indicators of governance maturity:

- Mentions of audits, compliance, or regulators

- References to global teams or external partners

- Language around accountability or data ownership

These cues matter. They usually signal that access decisions must be explainable, traceable, and reviewable—not merely functional.

Do Not Answer As an Admin Under Time Pressure

Many experienced admins struggle with this exam not because they lack knowledge, but because they answer as if they are fixing an urgent ticket.

The exam expects you to slow down.

Before committing to an answer, ask

- Would I design this knowingly, or am I reacting tactically?

- Would I document this decision with confidence?

- Would I defend this design in front of security, legal, and leadership?

If the answer feels rushed, it is likely incorrect.

Use Elimination Strategically

Rarely are all four options equally plausible. Eliminate answers that

- Hard-code assumptions

- Ignore future scale

- Create hidden access paths

- Require manual cleanup without controls

Often, the correct answer is not exciting—it is simply the least risky.

Trust Principles Over Platform Changes

Salesforce evolves rapidly, but security and sharing principles change slowly. Exams may update terminology or features, but foundational concepts—least privilege, clear ownership, separation of concerns—remain constant.

If you are unsure between two options, choose the one that aligns more closely with these principles rather than newer or flashier capabilities.

Practice Explaining "Why," Not Just "What"

The Security and Sharing exam increasingly tests architectural judgment rather than configuration recall. When reviewing any sharing decision, train yourself to articulate *why* one approach is preferred over another—especially in terms of scale, governance, auditability, and long-term maintenance. If you cannot explain the rationale behind a choice, you are likely relying on memorization rather than understanding.

Treat Every Question As a Design Review, Not a Quiz

Approach exam scenarios as if you were sitting in an architecture review board. Assume incomplete information, competing priorities, and future growth. Eliminate answers that solve the immediate problem but introduce risk over time. The correct option is often the one that balances security, performance, and adaptability—not the one that simply "works today."

Closing Thoughts

Security and sharing are often treated as constraints—necessary limitations placed on an otherwise flexible platform. But viewed through an architect's lens, they are something far more important—expressions of trust, accountability, and organizational maturity.

Every access decision tells a story. It reveals how an organization values data, how it balances transparency with responsibility, and how thoughtfully it prepares for change. The strongest sharing models are not the most complex or the most locked down; they are the ones that remain understandable long after their original designers have moved on.

As you close this book, resist the urge to see sharing as "finished." Instead, see it as a living architecture—one that deserves regular reflection, deliberate evolution, and occasional simplification. Whether you are preparing for an exam, designing a new implementation, or re-evaluating an old one, the real measure of success is not whether access works today, but whether it still makes sense tomorrow.

If this book has helped you ask better questions about visibility, ownership, and intent, then it has served its purpose.

Index

© Sandhya Sharma 2026
S. Sharma, *Unlocking Salesforce Data*, https://doi.org/10.1007/979-8-8688-2305-3

F

Field-level security (FLS), 3, 12, 13, 16, 17, 19, 20, 22, 29, 31, 56, 85, 213

Field Technician profile, 36
Fine-grain access control, 47, 49
FOIA reviews, 287
Full Access permission, 235

G

General Data Protection Regulation (GDPR), 1, 47
"Good enough" sharing model, 228
Governance

Granting access, 138

H

Health Insurance Portability and Accountability Act (HIPAA), 1, 47
Healthcare organization, 50
HIPAA, *see* Health Insurance Portability and Accountability Act (HIPAA)

I

Implicit sharing, 241, 242
Inherited sharing models, 279
Institutional knowledge, 202
Intentional segmentation, 201
Investigative access, 164

J

Jitendra Zaa, 214–215

K

KISS principle ("Keep it Simple, Stupid!"), 217
Knowledge continuity, 202

L

Layered access, 211
"Least privilege", 278

GPSR Compliance
The European Union's (EU) General Product Safety Regulation (GPSR) is a set
of rules that requires consumer products to be safe and our obligations to
ensure this.

If you have any concerns about our products, you can contact us on

ProductSafety@springernature.com

In case Publisher is established outside the EU, the EU authorized
representative is:

Springer Nature Customer Service Center GmbH
Europaplatz 3
69115 Heidelberg, Germany

www.ingramcontent.com/pod-product-compliance
Lightning Source LLC
Chambersburg PA
CBHW080859160726
48000CB00009B/2774